T0141877

Science Policy Reports

The series Science Policy Reports presents the endorsed results of important studies in basic and applied areas of science and technology. They include, to give just a few examples: panel reports exploring the practical and economic feasibility of a new technology; R&D studies of development opportunties for particular materials, devices or other inventions; reports by responsible bodies on technology standardization in developing branches of industry. Sponsored typically by large organizations—government agencies, watchdogs, funding bodies, standards institutes, international consortia—the studies selected for Science Policy Reports will disseminate carefully compiled information, detailed data and in-depth analysis to a wide audience. They will bring out implications of scientific discoveries and technologies in societal, cultural, environmental, political and/or commercial contexts and will enable interested parties to take advantage of new opportunities and exploit on-going development processes to the full.

More information about this series at http://www.springer.com/series/8882

Hans Peter Beck • Panagiotis Charitos

Editors

The Economics of Big Science

Essays by Leading Scientists
and Policymakers

Foreword by Rolf-Dieter Heuer

 Springer

Editors
Hans Peter Beck
Albert Einstein Center for Fundamental
Physics, Laboratory of High
Energy Physics
University of Bern
Bern, Switzerland

Panagiotis Charitos
Department of Experimental Physics
CERN
Geneva, Switzerland

ISSN 2213-1965   ISSN 2213-1973   (electronic)
Science Policy Reports
ISBN 978-3-030-52393-0   ISBN 978-3-030-52391-6   (eBook)
https://doi.org/10.1007/978-3-030-52391-6

This book is an open access publication.

Cover illustration: Photo composition by Teo Kontaxis. Image credits CERN, ESA

This Springer imprint is published by the registered company Springer Nature Switzerland AG.
The registered company address is: Gewerbestrasse 11, 6330 Cham, Switzerland

Foreword

A price worth paying.

Science, from the immutable logic of its mathematical underpinnings to the more fluid realms of the social sciences, is one of the most important drivers of economic progress. In its purest form, science generates knowledge. The applied disciplines translate that knowledge into tangible benefits for society, generating new tools for basic research along the way in a constantly evolving virtuous circle. The social sciences help us to make sense of the whole process and to understand how science is conducted, deployed, and perceived.

Science has carried us from our humble origins to an understanding of such esoteric notions as gravitation and quantum mechanics. It has applied this knowledge to develop devices such as GPS trackers and smartphones. And it allows us to model how such advances influence us and in turn the world we inhabit.

Science is an extremely powerful undertaking and one that is not always linear. No one can deny that science has led us up blind alleys, or taken wrong turnings. Yet its strength lies in its ability to process data, to self-correct, and to form choices based on the best available evidence. Thanks to this, it is equally undeniable that science has led us to a better world than that inhabited by our ancestors and that it will continue to deliver intellectual, utilitarian, and economic progress.

Last year's workshop on the Economics of Science, organized in the context of the 2019 FCC week, provided an opportunity to take stock of the interconnectedness of science and economics, with talks covering aspects ranging from procurement to knowledge transfer, and from global impact assessment to specific case studies. Its conclusions represent an important contribution to the literature and a guide for future projects.

In accelerator-based particle physics, the tools of the trade are highly sophisticated, large, and costly instruments. Providing an economic analysis for new projects is therefore something that the field has been accustomed to for a long time. CERN, for example, has been publishing economic impact assessments since the 1970s. As the scale of our endeavor grows, so does global collaboration. To find the last major facilities in particle physics that were built by single labs, we have to go back to the 1980s with LEP at CERN, the SLC at SLAC, and the Tevatron at Fermilab. Today, all potential future projects are planned globally, taking science, utility, and economics into account.

Of course, it is not only the labs that examine their economic and utilitarian impact. Society, quite rightly, holds a mirror up to science through independent reviews and media scrutiny. There are those who feel that limited resources for science should be deployed in areas such as addressing climate change, rather than in blue sky research. These views can be persuasive, but they are misleading. Science functions best as a whole. Blue sky research is every bit as important as directed research, and through the virtuous circle of science, they are mutually dependent.

In April 2020, as a curtain-raiser to the forthcoming update of the European Strategy for Particle Physics, Nature Physics published a series of articles about potential future directions for CERN. The journal's editorial pointed out that the scientific case for the future of particle physics is strong and the utilitarian argument compelling. "Even if the associated price tag may seem high—roughly as high as that of the Tokyo Olympic games," the editorial reads, "it is one worth paying." This report, along with the hard work that has been put into making an integrated scientific, utilitarian, and economic case, provides valuable evidence in support of this conclusion.

Former Director General, CERN,   Rolf Heuer
European Organization for Nuclear
Research, Geneva, Switzerland

Hamburg University, Hamburg,
Germany

Contents

Introduction

Hans Peter Beck and Panagiotis Charitos

The present volume collects the proceedings of the workshop "The Economics of Science" that was held in June 2019 in Brussels in the framework of the Future Circular Collider (FCC) Week with the support of the H2020 EuroCirCol and EASITrain projects and the Belgium Charter of the LSE Alumni Association. The goal of the meeting was threefold: First to explore the role of public investments in research infrastructures and Big Science projects for economic development, review ways to access their financial impact beyond their core scientific mission and thirdly create a forum for exchanging best practices that can maximize the impact of such projects. The collected essays focus on Big Science Organizations that participated in the workshop while we should clarify to readers that by "Science" we mainly refer to curiosity-driven research. However, we hope that some of the ideas and tools discussed by the participants of the workshop can find applications in many ways.

The economic and social benefits of Research & Innovation don't happen by magic; they often have to start with curiosity-driven research, not directed to applications but to explore the nature of our universe and our place in it. What is much less well known is the wider impact this has on technology and our daily lives; fundamental, exploratory science that poses high-risks but also delivers surprising results in tackling some of the most pressing societal problems and unlocking new markets potential. A mere 150 years ago the candle was the main source of artificial light. By now, lighting has been developed to a very sophisticated degree. In Oren Harari's famous quote: "The electric light did not come from the continuous improvement of candles". No amount of research on the candle would have given

H. P. Beck
Albert Einstein Center for Fundamental Physics, Laboratory of High Energy Physics,
University of Bern, Bern, Switzerland
e-mail: Hans.Peter.Beck@cern.ch

P. Charitos (✉)
Department of Experimental Physics, CERN, Geneva, Switzerland
e-mail: panagiotis.charitos@cern.ch

© The Author(s) 2021
H. P. Beck, P. Charitos (eds.), *The Economics of Big Science*, Science Policy
Reports, https://doi.org/10.1007/978-3-030-52391-6_1

us the electric light bulb which was only made possible through basic science that unveiled the nature of electricity and gave birth to numerous applications. Another example is the phonograph that Edison invented in 1877 based on making bumps in a metal surface and turning them into sound by running a mechanical "finger" along them. Despite years of improvements in the material, development of better bearing and support structures, the real revolution in our listening experience came with the development of MP3 technology. It took a courageous step followed by a long development process to store sound in a digital format thus revolutionizing the quality and volume of sound we can store today. **In conclusion, only focusing on small step improvements that seemingly will lead to the next iteration ready to market, big opportunities will be missed that for short or long will turn out detrimental in any business model, if not in parallel basic and fundamental research are maintained at a healthy level and opportunities that open up from it are embarked on.**

The unprecedented pace of scientific discoveries during the 18th and 19th century that also led to the Industrial Revolution went hand in hand with the development of economics as a separate discipline; developing its own tools and methodology and advancing taking into account the progress of other fields from psychology and sociology to mathematics and computing. However it is our belief that today the pendulum is swinging back and this has been one of the motivations for preparing this volume. The contributions in this publication demonstrate, economists and scientists are coming closer together to realise the strong links between basic research and its societal impact. Today, research facilities, academic institutes, private industry and funding agencies embrace increased multi- and transdisciplinary research to tackle the world's most challenging problems.

A half-century ago, Gordon Moore wrote a paper in which he projected that progress in the density and speed of silicon chips would increase exponentially. In his paper, Moore envisioned how this would enable technologies ranging from the personal computer, to the smartphone, to the self-driving car.[1] His prediction became known as Moore's Law, and it has held remarkably true for 50 years. At the celebration of the fiftieth anniversary of his seminal paper back in 2015, Moore talked about the impact of his insight on modern technology and the crucial role of basic scientific research for realizing it. In his own words: "That's really where these ideas get started. They take a long time to germinate, but eventually they lead to some marvelous advances. Certainly, our whole industry came out of some of the early understanding of the quantum mechanics of some of the materials. I look at what's happening in the biological area, which is the result of looking more detailed at the way life works, looking at the structure of the genes and one thing and another.

---

[1]In Moore's words: "The future of integrated electronics is the future of electronics itself. The advantages of integration will bring about a proliferation of electronics, pushing this science into many new areas. Integrated circuits will lead to such wonders as home computers—or at least terminals connected to a central computer—automatic controls for automobiles, and personal portable communications equipment." Ref: https://newsroom.intel.com/wp-content/uploads/sites/11/2018/05/moores-law-electronics.pdf

These are all practical applications that are coming out of some very fundamental research."

The technological progress we have enjoyed over the last century was enabled by the combination of science education and investments in fundamental research. While the opportunities for discovery have never been greater, commitment to and funding for basic science seems to be put under question. It is often seen as absorbing sources from rather more target-oriented research to address major issues that affect our everyday lives such as climate change, infectious diseases, cancer therapies, natural hazards. However this view neglects that although basic science might not offer clear-cut ways to immediately solve problems, it is the bedrock for future fixes. However, these are not "either/or" options but rather based on a "both/and" framework that better describes the positive symbiosis between different fields.

This is why, in our view, it is urgent to foster the dialogue between different actors including researchers, funding agencies, policy makers, economists and innovators among others. We need to learn from previous lessons, exchange current best practises and develop synergies across big scientific organizations that act as hubs for excellent science and catalyzers of global collaboration.

Basic research occurs at a distance from the market, which makes it unappealing as in our society we are more and more used to acting in short- to medium-term timeframes. Once a new idea becomes clear to be useful it still needs a lot of energy and dedication to turn into reality and be used to serve society. The history of science offers ample examples on how the benefits of basic research can take up to decades before translating into innovations and generating a positive return for society. Furthermore, science, and particularly basic science, is also inherently unpredictable about its results, bearing risks and calling for a broader vision that drives scientific progress. Failure is common but success in understanding our world is unmeasurable. How could one quantify the value of discovering X-rays, the study of synchrotron radiation and the fundamental laws of optics about diffraction that played a crucial role from the first moment in uncovering the structure of DNA and the development of new generations of medical therapies? In addition, one should not undermine the value of often unexpected discoveries, the so-called "unknown-unknowns", like the discovery of the Lamb shift paved the way to Quantum Electro Dynamics (QED) or the surprising observation of spin by Uhlenbeck and Goudsmit in fundamental particles that paved the way for understanding material properties and are crucial when developing new materials with distinct wanted features. Basic research on spin made MRI imaging possible, now widely used in medical diagnostics.

The knowledge generated through curiosity-driven research often disperses into the wider economy as a shared public good thus making it harder to track and quantify the generated revenue. For example in the recent fight against COVID-19 how could one quantify the role of understanding the behavior of complex biological molecules and fundamental research in biology or the role of particle accelerators and electron free lasers in understanding the structures of various aspects of the CoV-2 spike protein through crystallography and electron cryo-microscopy (cryo-EM) and getting valuable information for the development of treatment therapies.

The above mentioned reasons reflect to a certain exchange the challenge of mapping the wider economic impact of curiosity-driven science. Significant efforts in the last decades already shed light to some of the issues tight to this relation. Perhaps at this point we should pause for a moment to reflect on the value that curiosity driven research carries by itself. It is par excellence the field that brings together creative minds to collaborate on curiosity driven research; encouraged to take risks but also be prepared to fail. This is not to undermine the value of more applied-oriented research that transforms these ideas into applications that can benefit society at large. In fact basic and applied research become more intertwined as we are entering the realm of Big Science calling for a broader view. On this note we would like to repeat that most of the essays collected in this volume don't discuss the direct scientific output of this research - since this is and should remain risk-free following the scientific method and judged by the different scientific communities— but rather trace additional impacts generated through the investment in curiosity-driven scientific efforts.

The contributions in this volume present work done by various research infra-structures and big scientific projects including the European Organization for Nuclear Research (CERN), the European Space Agency (ESA), the European Spallation Source (ESS) and the proposed Square Kilometer Array (SKA). A number of them summarize how these big scientific organizations try to measure the impact that they generate for the society and economy beyond the core scientific questions that these large-scale infrastructures try to answer.

Simon Berry (SKA's Director of Corporate Strategy) discusses the SKA's approach in creating a sustainable research infrastructure, well-embedded in local networks and scientific communities while also attracting users from all over the world and generating societal benefits. John Womersley (Director General of the European Spallation Source) outlines some of the key challenges in building sus-tainable support for any science megaproject, using the European Spallation Source ESS as an example. Thierry Lagrance (CERN, Head of Industry, Procurement and Knowledge Transfer) reflects on CERN's approach in accelerating and optimising the generation of socio-economic benefits. Charlotte Mathieu (ESA's Head of the Industrial Policy and Economic Analysis Section) presented the existing framework for assessing the impact of ESA programmes and key lessons derived from past consultations. Finally, Philip Amison (Head of Corporate Strategy and Impact Science at STFC/UKRI) reviews the key findings of an evaluation of the benefits that the UK has derived from CERN based on a 2018 commissioned by STFC and performed by Technopolis. The report captures and measures the range of scientific, economic and social impacts emerging over the past decade, considering both direct UK involvement as well as the wider impact that CERN has on the UK.

Moreover, in this volume, economists and policy makers present the meaning of sustainability in the context of large-scale research infrastructures and some of the existing tools that can inform empirical studies of RI's socio-economic impact. Margarida Ribeiro (European Commission, Directorate for Research & Innovation) presents the complex and often multi-level sustainability challenges for Research Infrastructures and the key ingredients of a coordinated plan for action among

European RIs. Alasdair Reid (Policy Director, European Future Innovation System—EFIS Centre) in his essay offers an overview of the main outcomes of the H2020 RI-PATHS project that aims to provide policy makers, funders and RI managers the tools to assess RI impact on the economy and their contribution to resolving societal challenges. Andrea Bastianin (Ass. Professor, University of Milan—Bicocca) summarizes the results of a social Cost–Benefit Analysis (CBA) of the High Luminosity upgrade of the Large Hadron Collider (HL-LHC) and the merits of this method when applied on Research Infrastructures. His talk followed the presentation of the merits and basic ingredients of a CBA method by Massimo Florio (Professor, University of Milan) arguing that CERN and more generally Big Science Centres (BSCs) are ideal testing grounds for theoretical and empirical economic models while demonstrating the positive net impact that the LHC/HL-LHC has for society. Moreover, Silvia Vignetti (Director, CSIL) in her contribution suggests that to inform strategic planning, the data collection for impact assessment should not be episodic and motivated by external requests from stakeholders and funding agencies but rather a well-integrated activity occurring throughout the lifetime of the infrastructure.

The socio-economic impact of research infrastructures extends over longer time and spatial scales. In her essay, Linn Kretzschman (MSCA ESR, University of Vienna) presented some of the ongoing research on how society can benefit from a new research infrastructure during the design, construction and operation phase. Her work in the Institute of Entrepreneurship and Innovation at Vienna University of Economics (WUW), supported by the H2020 EASItrain project, has identified innovative application fields, outside particle physics, for the required superconducting magnets for a next-generation collider. To identify new market opportunities for superconducting magnets and its manufacturing steps, the team analyzed the full manufacturing value chain with regard to their importance and identified numerous opportunities that were further scrutinized for their research potential. Riccardo Crescenzi (Professor, London School of Economics) discussed how RIs potential for innovation increases when coupled with complementary skills and conditions are available locally to support knowledge generation and absorption. Investments in R&D can enhance regional innovation only when coupled with a supportive endowment of Human Capital. Finally, Maria L. Loureiro (Professor U. Santiago de Compostela, Spain) and Maria Alló (U. Santiago de Compostela, Spain) invite us to rethink the concept of value, how it is defined and offer a more inclusive approach that is appropriate when dealing with global public goods like those created by Big Scientific Infrastructures.

The last part of the volume brings together three essays focusing on the question "Who benefits from such large public investments in science" that informed a public discussion moderated by Mrs. Anjana Ahuja during the last session of the workshop. During the session Massimo Florio (Professor, University of Milan) introduced the question of how Big Science contributes to social justice and can contribute in tackling current inequalities. Two fundamental questions are addressed: What is the economic impact of curiosity-driven research? What are the implications for social justice of the interplay between—on one side—government funded science and—on the other side—R&D supported by business? Arguing for the need to include this dimension on top of discussions about the size of public funding and the socio-economic impact they create. Michela Massimi (Professor, University of Edinburgh) offers her remarks on the importance of fundamental research for society and how it contributes to the human cultural flourishing while arguing that philosophers of science should contribute more and more in the decade to come in this ongoing dialogue and engagement with physicists. Finally, Carsten Welsch (Professor, Head of Physics Department University of Liverpool/Cockcroft Institute) argues that fundamental science informs many aspects of our daily lives and should not be considered as a distant activity and demonstrates that through an extensive discussion of the development and applications that particle accelerators have with more than 50,000 particle accelerators used in industry, for medical treatment and for research.

Closing this editorial we would like to refer to the special impact that research infrastructures for curiosity-driven research have for training the next generation of science. Acting as knowledge and innovation hubs they offer an international and competitive environment, characterized by creativity, collaboration and resilience is key for succeeding in their transition from curiosity-driven research to different sectors where they often contribute in unprecedented ways. The next scientific revolution will be driven by scientists who have a multidisciplinary view of science, the opportunity to take risks, the infrastructure to work, and the freedom to think.

We hope that the publication of these proceedings will inform the greater debate on the value of public and private funding for research infrastructures and inform discussions on the broader value that public investment in curiosity-driven science and research infrastructures for a transition to more resilient and redistributive model of economy in line with the big societal challenges lying ahead in the twenty-first century.

Towards a Sustainable European Research Infrastructures Ecosystem

Margarida Ribeiro

Contents

High-quality, accessible research infrastructures (RIs) are at the heart of the knowledge triangle of research, education and innovation. Whether operated and funded at regional, national or transnational level, they play a key role in advancing, exploiting and disseminating knowledge and technology while facilitating cross-sectoral international collaboration.

The European research and innovation infrastructure ecosystem is diverse in scale and scope, comprising numerous facilities and stakeholders operating at the leading edge of scientific discoveries. Moreover it provides hands-on support to local or regional stakeholder communities. Through these initiatives, the EC supports tens of thousands of researchers in academia and industry to develop innovative ideas, products and services that foster European competitiveness and help tackle some of the most pressing societal challenges that we face today.

Ensuring the availability of state-of-the-art facilities requires multi-billion Euro long-term investments across Europe. The funds come mainly from different national funding instruments. EU's Framework Programme for R&I which amounts to a total of 4.1 billion EU (1.7 billions under FP7 and 2.4 billions under H2020) is a smaller but still useful fraction. Moreover, EU's Developments and Cohesion funds have also contributed to the developments of European RIs with a total amount of 18.2 billion EU in the last two programme periods (11.6 in 2007–2013 and 6.6 in 2014–2020).

Public investment in RI's is justified given the crucial and multi-faceted role they have in advancing our knowledge around certain scientific fields, training the next generation of scientists, engineers and supporting R&D actions that enhances Europe's innovation capacity.

M. Ribeiro (✉)
EU, DG for Research & Innovation, Brussels, Belgium
e-mail: Margarida.RIBEIRO@ec.europa.eu

© The Author(s) 2021
H. P. Beck, P. Charitos (eds.), *The Economics of Big Science*, Science Policy Reports, https://doi.org/10.1007/978-3-030-52391-6_2

Given their multiple roles, it is important to develop a modern, effective and sustainable RI ecosystem that will enable Europe to maintain its leading role in an increasingly globalized and competitive environment. For this to happen, we need to coherently integrate new infrastructures into the European landscape, plan for international governance structure, leverage the capacity to serve an international community of users and finally develop adequate long-term funding for the construction and operation, while striving to maximise their societal benefits.

This set of needs also raises the challenge to agree on the methodology for assessing RI's impact and ensure the long-term and stable commitment of different actors involved in the different phases of a RI's life-time.

The importance of ensuring the long-term sustainability of RIs was stressed on a number of occasions. It had already been flagged as a policy priority in the informal Competitiveness Council of July 2014. More recently, as a result of the May 2016 Competitiveness Council Conclusions [1], the Commission was invited to develop an RI long-term sustainability Action Plan, in close cooperation with ESFRI and other relevant stakeholders. The input of a stakeholder consultation process culminated in the publication of the Commission Staff Working Document "Sustainable European Research Infrastructures—A call for action" [2].

The key goals of the consultation were:

- Ensuring scientific excellence
- Attracting and training the managers, operators and users of tomorrow
- Unlocking the innovation potential of RI, Measuring socio-economic impact of RI
- Exploiting better the data generated by the RI
- Establishing adequate framework conditions for effective governance and sustainable long-term funding for RI at every stage in their life-cycle
- Structuring the international outreach of RI.

The following scheme summarizes some of the emerging themes that were described in the EC report on the Long-Term Sustainability of RIs.

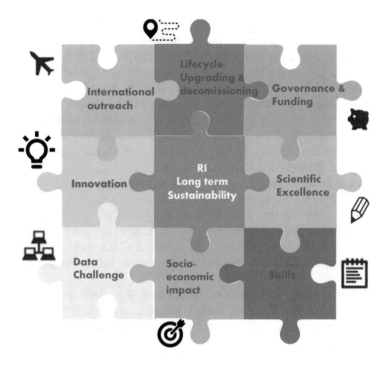

Key priority is to **ensure excellence of the services provided by the RIs.** Today it is widely accepted that excellence is the main driver for the development of RIs and this should be supported through the entire lifetime of RIs including the pursuit of research as well as the development of new technologies for and by the RI's users. Most of the stakeholders, among which ESFRI, indicated in this respect a need to develop guidelines for standardized, effective and robust evaluation procedures of RI through independent international peer-review as an active measure to increase the widespread adoption of such instruments[1] [3]. Knowledge, education, technology, and innovation are Europe's main strengths and the foundation for growth and employment thus when discussing the sustainability of RIs we need to think how RIs contribute to these areas.

Another important aspect is the human capital formation that profits from RIs ecosystem. This includes the users, the managers and the operators (among other categories)[2] [4]. When designing future RIs we need to think of the different

---

[1] See ESFRI long-term sustainability WG report: https://ec.europa.eu/research/infrastructures/pdf/esfri/publications/esfri_scripta_vol2.pdf

[2] See also the EC's charter for access to research infrastructures which sets out principles and guidelines when defining access policies. https://ec.europa.eu/info/files/charter-access_en

categories who are likely to be involved in different capacities throughout their lifetime. Therefore, a challenge and a take-home message is that the right people should be in the right place and at the right time. Today about 53% of the European RI do not apply international peer review for the selection of the user projects and for attributing access; the situation needs to be rapidly addressed for present and future RIs. A way to tackle this issue would be to include a defined access policy of a RI as a requirement for funding or develop an access system based on excellency criteria. Moreover, about 21% of the RIs do not have in place an International Advisory Scientific Committee leading to the proposal that RIs should establish Technical Evaluation and Management Assessment committees.

Excellence in research requires top infrastructures for data collection, management, processing, analysing, and archiving. We need to think how we can better exploit the data from RIs and make them accessible to a wider community of users (open data, open analysis, preservation are among themes that enter this discussion). Finally, one should not underestimate the potential of RIs as hubs of innovation through the support of an ecosystem between researchers, entrepreneurs and industry around RIs.

Furthermore, one of our priorities is to encourage RIs to act as early adopters of technology and promote R&D partnerships with industry to facilitate industrial use of research infrastructures and stimulate the creation of innovation clusters. The goal is to make industry more aware of the opportunities to improve their products and co-develop technologies needed to proceed in fundamental research that could also have a market potential. Our aim is to implement policies that encourage young researchers to develop their ideas and provide the required resources that will allow them to think how these ideas can be transformed into marketable products and services. In this respect, support for RIs as hubs of interdisciplinary collaborative research in new and promising fields, can lead to greater European competitiveness, employment, and prosperity.

Two further challenges, that I would like to share with you are related to the assessment of the economic and wider societal value of RIs and the establishment of adequate framework conditions for effective governance. The first calls for the development of a standardised model, using certain data input and key indicators for identifying the socio-economic impact of RIs. A topic that also informed the organization of this workshop. As highlighted by ESFRI, national authorities and funding bodies should be explicit about the role that socio-economic impact plays in their strategy and funding decisions. Clarifying this role of RI's will help operators and users to take appropriate action when developing future strategic plans and operating models. In that sense, periodic monitoring of societal impact should be an integral part of the regular assessment of RIs over its whole lifetime. In addition, discussions with stakeholders indicated the need to better assess the intangible investments, in quantitative terms, as today they remain rather poorly understood. This is one of the main goals of the H2020 "RI-PATHS: Research Infrastructure imPact Assessment paTHwayS" project; namely to develop a consistent, empirically implementable and holistic model at international level for analysing the socio-economic impact of research infrastructures and their related financial investments.

The second challenge is related to the need of synchronising national roadmaps with the newly established ESFRI and ERICs. Funding will be needed to support not only the construction and operation but also the transnational and virtual access of researchers and the harmonisation and improvement of the services that the infrastructures provide while an important factor that should not be neglected are the costs associated to the decommissioning of an RI at the end of its lifetime.

Finally, we need to strengthen the international dimension of pan-European RIs increasing their visibility at global scale and paving new ways for cooperation. The global dimension of societal challenges but also the complexity of new tools to push the frontiers of human understanding of the fundamental laws of nature call for better coordination and collaboration at a global level. In fundamental science, projects like the LHC that lead to the discovery of the Higgs boson or the Event Horizon Telescope that gave us the first image of a black hole clearly demonstrate this need that inform the design of future RIs.[3] The scientific challenges addressed by European research infrastructures makes increasingly relevant their cooperation at European but also international level while exploiting synergies with research infrastructures in other world regions as well as the development of global research infrastructures.

The Research Infrastructure Programme within EU's Horizon Europe framework for Research & Innovation aims to address these challenges by (but not only):

1. Consolidating the Landscape of European Research Infrastructures
2. Opening, Integrating and Interconnecting Research Infrastructures
3. Reinforcing European Research Infrastructure policy and International Cooperation

Each of these items breaks down into a number of specific lines of actions. I will summarize some of the key points below while for more information you can visit the EU pages for Research & Innovation [5].

Regarding the consolidation of new RIs it is important to clearly define the structure for each of the design, preparatory and implementation phases, with clear and different criteria appropriate to these levels. This will help to identify the complementarity between different funding resources while also facilitating service agreements between RIs giving the option for upgrade, merging or decommissioning of RIs.

---

[3]Note of the Editor: this is also clearly demonstrated in the global fight against COVID-19 where numerous RIs helped to identify the structure of the new virus and in the development of new medical treatment along with the required computing resources and supercomputers for simulating the response to possible therapies as well as for tracking the evolution of the pandemic and design public health measures. A full list of EU-funded projects supporting research to tackle the COVID-19 pandemic can be found here: https://ec.europa.eu/info/research-and-innovation/research-area/health-research-and-innovation/coronavirus-research_en

Furthermore, it remains imperative to consider the scalability and sustainability of the European Open Science Cloud (EOSC) bringing together European, national, regional and institutional resources. EOSC's evolution should take into account present and future needs from different research communities and the emerging technologies that could address them. This should facilitate the preservation and open access to data from RIs in compliance with the FAIR principle. Finally, in the same direction any future RI should support and be well-integrated in a pan-European research and education network.

The second pillar refers to the preparation, implementation, long-term sustainability and efficient operation of the research infrastructures identified by ESFRI and of other world-class research infrastructures, which will help Europe to respond to grand challenges in science, industry and society. The evolution of the transnational research facilities implies that RIs become elements of "supra-national innovation systems" and, in this setting, industrial players can play the role of potential supplier (of the required technologies), user and co-developer. As discussed, we need to put in place appropriate governance models and foresee a framework that will provide access of services between different RIs at regional, national and EU level. Finally, RIs managers would benefit from the development of common strategic roadmaps on the required R&D to advance certain technologies and help improve their services through partnership with industry. The innovation potential of RIs can also be expressed through the development of new services. RIs can also trigger new business models and services to policy makers.

A consultation among stakeholders has revealed the need to:

- Increase RI engagement with industry, by fostering their direct and early involvement in Advisory Boards.
- Enhance the role of intermediaries (ILO's) and develop specific mechanisms to stimulate the commercial application of RI services and tools.
- Clarify apriori industry access rules, mainly concerning IPR regimes and procedures for accessing RI.
- Stimulate joint innovative procurement mechanisms, pre-commercial procurement and the link with Public Procurement of Innovative Solutions.
- Develop strategic roadmaps in key technologies required for the construction and upgrades of RI in close relation with EIT, KICs and KETs.

Therefore, a co-creation approach to continuously generate, scale and deploy breakthrough technologies with market and social value remains a big challenge regarding European Research & Innovation landscape.

Currently the new Research Infrastructure Programme of Horizon Europe is under development. It is among our priorities to improve synergies with other EC programmes while coordinating a Pan-European ecosystem of RIs. I believe that synergies between different funding instruments could strengthen the position of European RIs in the innovation ecosystem at EU level and to users developing new markets for key technologies.

References

1. Report on the Consultation on Long Term Sustainability of Research Infrastructures May 2016. EUROPEAN COMMISSION Directorate-General for Research and Innovation Directorate B—Open Innovation and Open Science: https://ec.europa.eu/info/sites/info/files/research_and_inno vation/research_by_area/documents/lts_report_062016_final.pdf
2. Sustainable European research infrastructures, A call for action: Commission staff working document: long-term sustainability of research infrastructures (2017) [DOI: https://doi.org/10. 2777/76269]: https://op.europa.eu/en/publication-detail/-/publication/16ab984e-b543-11e7-837e-01aa75ed71a1/
3. ESFRI long-term sustainability WG report: https://ec.europa.eu/research/infrastructures/pdf/esfri/publications/esfri_scripta_vol2.pdf
4. EC's charter for access to research infrastructures which sets out principles and guidelines when defining access policies. https://ec.europa.eu/info/files/charter-access_en
5. https://ec.europa.eu/info/departments/research-and-innovation_en

Full Presentation

https://indico.cern.ch/event/727555/contributions/3461262/attachments/1867872/3072630/RI-Long_term_sustainability_FCC_week_MR.pdf

Disclaimer

"The views represented in this paper are personal and do not represent those institutions or organisations that the author is associated with in her professional capacity unless explicitly stated. The European Union cannot be held responsible for any use which may be made of the information contained therein."

Economics of Science in the Time of Data Economy and Gigabit Society

Michal Boni

In a period that Europe prepares to launch its new multi-annual programme for research and innovation it is important to step back and reflect on the broader social and economic benefits that Big Science and curiosity-driven research have brought to society and moreover how we can work together to maximize this impact. Further engaging citizens, offering more opportunities for young people, exploiting the tools presented by digital technologies are some of the challenges that I highlight/discuss in this essay.

The possibilities to build a real ecosystem of knowledge economy rely on many factors, which need to be seen together. It is the concept of the knowledge society development, the institutional and financial background for the Research and Innovation (R&I) growth and the unbelievable data economy increase over the last years. All of them have contributed to the re-designed model of the knowledge economy. Science and innovation is now in the heart of the future development and is building competitive advantages of the European Union in the mid and long-term perspective.

In 2018, the **innovation** becomes the crucial reference point to establish one of the key growth engine for the EU under the next Multiannual Financial Framework. The proposed Horizon Europe for 2021–2027 was equipped by the European Commission with nearly EUR 100 billion. This was fully supported by the European Parliament, even if some wished to allocate more, ca. EUR 120 billion.[1]

---

[1] In the new document presented on May 2020 by the European Commission the position of the Horizon Europe is very strong, with the similar expenditures, and with some reorientations addressed to the new post-Covid challenges, As R&I role in building the sustainable, inclusive and resilient recovery, in addition on virology, vaccines development, treatment and diagnostics, also acceleration of the twin digital and green transition (COM (2020) 442 final),The EU budget powering the recovery plan for Europe, May 2020.

M. Boni (✉)
University of Social Science and Humanities (SWPS University), Warsaw, Poland

© The Author(s) 2021
H. P. Beck, P. Charitos (eds.), *The Economics of Big Science*, Science Policy Reports, https://doi.org/10.1007/978-3-030-52391-6_3

But it is not only financial envelope that matters. What is equally essential is the establishment of new rules on how to use the resources dedicated to the Science in practice.

In this respect, the fundamental issue was how to strengthen the cooperation between businesses and the science. What kind of conditions are necessary to develop innovative businesses, to inspire the new innovative spirit among scientists, to empower the collaborative forms helping both partners finding the most adequate and effective solutions. An improved **partnership** was a right solution, as it was oriented at various mutualities, supportive to achieve diverse advantages, but also to share burdens and concerns, as well as jointly solve problems. All of this with a strict requirement of the most appropriate and efficient leadership of business/scientific projects.

It does not mean, however, that only projects with a prospect of a success should be considered for partners' efforts. The magic of science very often means: unknown territory, unpredictability and yet the brave readiness to explore completely new areas. Hence, a certain 'right to fail' should be incorporated into the costs of the new system as it already exists within the start-ups.

This is the only way to go forward with innovative ideas, disruptive inventions and research and enhance the possibilities for **commercialisation** of the results of research. Future commercialisation must be at the birth of any research and innovation project. Over last years, in many research centres and universities the business/ science relations were understood as a new challenge and established as a brokering model, often in the form of special spinoffs.

Taking into account this critical juncture and considering the opportunity for the **tipping point**, I would like to describe a general framework needed for the **Economics of Science**, especially in the time of data economy and gigabit society.

The following requirements are necessary to be fulfilled.

Firstly, what kind of infrastructure is crucial for science development in the modern times?

The most expected and needed is well-designed **data infrastructure**. There are some scientifically oriented dimensions of the data infrastructure.

First of all, the network should be able to deliver a high quality transmission of data, especially Big Data with regard to certain areas, such as genetic sequences, biotechnology collections of data to mention but a few. As the high quality of infrastructure depends on the proper correlation between the speed and the latency, the most promising solution is the 5G infrastructure development in the European Union based—step by step—on the Giga frequencies. It requires substantial investments but they are indispensable for the future oriented data based science.

The next dimension relates to a very specific infrastructural challenge enabling appropriate data processing. The development and the accessibility of the Super-Computers with exascale possibilities will play crucial role in the coming years. Just as the open access to the quantum technologies makes the analytics much more timely and adjusted to the modern science. There is no possibility to develop data transmission and processing without the European Network of High Performance Computing Centres (EuroHPCinitiative). The "network" means that all processed data can flow and some works and tasks can be shifted and shared (in the framework of cooperation and shared responsibility) from one HPCC to another. The added value for the science is pivotal. Moreover, there are also economic efficiencies: the majority of the European HPC projects is generating financial returns; each euro invested on average provides in return EUR 867 in increased revenue/income and EUR 69 in profits.[2]

The following dimension relies on proper design of the data infrastructure. It should be based on the European Data Strategy, which covers many areas and was—after many years of debates—finally presented by the European Commission in February 2020.[3] The proposed solution will allow establishing concrete Data Spaces with their functionalities: from agriculture, via industrial data, to the European Health Data Space. For the scientific field, the concept and implementation of the European Open Science Cloud was established a bit earlier, in 2017. However, for the effective functioning of all those Clouds and Spaces, a high level of the European interoperability (semantic) is required. It is essential for open and commonly addressed accessibility of data.

To sum up, the development of the data infrastructure is paramount for the economics of science as it can be one of the basic features of the European strategic autonomy. What is more, the digital components sovereignty with regard to the scientific research is necessary to ensure cybersecurity of all networks, including HPCC networks, SuperComputers and quantum technologies, and to provide for the necessary resilience against the digital scientific espionage.

Secondly, how can we use the **European Open Science Cloud** (EOSC) as an instrument to build and scale up the European Science?

Emblematically, the key word and practically, the key solution for the functionalities of the Science Cloud are based on the model of the **openness**. Presented by the European Commission, the triangle[4]: open innovation—open science—open to the world, was crucial for the public debate. Essentially, it was focused on the open mind and on the open society concept. In that sense, open science addresses the characteristic features of the modern society, oriented at the expected and developing the knowledge and innovative society. It is clear that such concept was easier to implement before the time of populism and the increasing role of the post-truth era.

The open access trend, crucial for the digital game changer in many areas (industrial, scientific, related to the entertainment and all kinds of public services, also important for everyday life) is increasingly significant for the science. To have an access to research articles, to various scientific papers, using different types of the accessibility from subscription to the Creative Commons (and licenses related to this model), is basic for the dissemination of knowledge. What is more, access to the results of the studies and to the meta-data shared information about format and the

---

[2]Developing supercomputers in Europe, European Parliamentary Research Service, October 2017, p.3.

[3]European Commission, A European Strategy for data (COM(2020) 66 final), February 2020.

[4]Open Innovation, Open Science, Open to the World, May 2016, with introduction by Carlos Moedas, the Commissioner for the Science.

content of the raw data, could further support the new scientific achievements. There is no possibility to be innovative without the broad access to many sources of information and knowledge.

But most importantly, the vehicle for the new science development is built on the significance of the Big Data, which at the same time is a fundament of the data driven economy.

Considerations around the Open Science Cloud model ought to take into account the importance of the rules for: data collecting, opening them for people (beneficiaries of the innovative solutions on one side and creators and scientists on the other), sharing, in some cases donating, processing, using, preserving and re-using, free flowing.

Also the context of the Open Science Cloud should be visible: The new legal framework for the free flow of data, harmonised implementation of the GDPR with a proper understanding of the meaning of "legitimate interest" and the use of "anonimisation" and "pseudonymization" schemes in the research, preparing for the future model of the ePrivacy directive, using new tools for the text and data mining under the Copyright Directive, openness for the accessibility of data for machine learning trainings, the re-use directive recast which is indispensable for genuine data exchange between Science and Research Centres (academia & universities) in order to overcome differentiations and limits established by Member States.

The key drivers of the European Open Science Cloud development are:

– Data culture and the culture of the data sharing: European science must be anchored in a common culture of data stewardship, so that research data is recognised as a significant output of research and is appropriately curated throughout and after the period conducting the research. It also means—as I expressed above—the re-use model of research,
– FAIR data governance: means—Findable, Accessible, Interoperable, Reusable. It should be built upon inclusive stakeholders participation. The various kinds of policies and raising the awareness of these principles should go hand in hand with technical implementation and social infrastructure, including education and training. For all partners, such as university associations, research organisations, research libraries, research data repositories and others like educational brokers, new skills and competences are needed. They could help people to live among datasets and to manage them.
– Data management plans which seems to be obligatory in all research projects generating or collecting publicly funded research data. The meaningful part of those plans should be addressed to avoid data fragmentation and unequal access to quality information sets. It is linked to the fragmented access across scientific domains, countries and governance models, limited cross-disciplinary access to data sets, and non-interoperable services. The challenge is how to allow for universal access to the scientific data (as a common good?) and how to establish an equal level playing field for the EU researchers and all kinds of users. We must build the European Open Science Cloud as a one stop shop to find, access and use

research data and services from multiple disciplines and platforms. Technically speaking, services and functionalities should be user driven.

It has taken three years to start legally and technically building the EOSC, and yet it has not been finalized. However, the developments are promising.

The problem is how to tie the European political will to make the research area much more open to fruitful exchanges and at the same time provide technical and infrastructural guarantee for effective—economically and scientifically—development.

Thirdly, the data economy is only one side of the economics of science. The other is gigabit society. The society, which is involved in all digital game changer achievements and concerns. In this context, what is the role of the Science, especially with regard to societal and social expectations, needs, threats and fears?

The fundamental question is whether the science development or the economics of science require the participation of the society. And this participation should not only be seen with regard to the receipt of the research results (it is obvious and exists as an essential, conventional model), but also as a possibility to create innovative solutions together with the science and scientists and to explain many scientific and societal problems. In the broad sense: to explain the world.

The above mentioned is especially crucial for a transparent and open communication and the **science literacy**.

Especially at the time of populism, when populistic emotions and narratives use fake news and fake science and attacking the open minded societies as well as critical thinking patterns. Populistic prejudices and stereotypes undermine the evidence and knowledge based policies and solutions. The genuine science is really damaged by "science-like" discourse that uses emotion-biased propaganda.

New social and moral obligations emerge. We need to reconsider the following key social aspects of the functioning of the science, which will immensely impact the economics of the science and its effectiveness:

- How to communicate about science, publicly and understandably,
- What kind of literacies are needed to establish a new relationship between science and society—it should be the digital literacy, but complemented and enhanced by democratic, social, scientific, media and political literacy for better understanding the environment of the modern science and the science as an innovative game changer,
- How to tackle the problem of emotions in research, preparing scientists through relevant trainings, using the collective intelligence,
- How to develop citizens' engagement in science, how to support and develop the public engagement. Public engagement is an ongoing process. One of the key recommendations should be how to understand the public and do not regard it as a single homogeneous mass, but diverse audiences.

Different media audience often include citizens who are organised in communities with a certain interest or support for a particular scientific field. Are they ready to participate in all kinds of scientific processes? By adding the citizens' value to these processes and supporting the scientific efforts by participatory models.

The **Citizens' Science** is not the utopian dream.

It can be very important to give people the feeling of **participation** in the scientific process. For example this include data collection about the environmental threats by measuring the air pollution as a responsible community, complementary to research analysis or measuring the water pollution or its deficits. Such initiatives have a twofold role: On one hand, it can be an expression of the visible societal behaviour (to save the Planet), on the other it can be a serious contribution to the scientific analysis.

This is how the citizens and the society could emphasize their views on many issues and create the so-called demand for the science in general terms: to take into consideration, to start deeper analysis, to observe processes and to find conclusions and recommendations. It is the phenomenon of the usefulness of scientific efforts and studies. In my view, it is a meaningful complementary contribution to the innovative science: involving representatives of all generations, but especially numerous young creative minds (pupils from schools and students from universities) to shape the future and raise a common awareness of the development based on the real evidence and future proof orientations.

Paramount is to drive the science development ecosystem change with the involvement of the citizens. It is crucial for the positive results of the economics of science: societal approval, opportunities for participation, understandability, openness for all kinds of contexts.

There is one more meaningful context of this change: the AI development—as a General Purpose Technology, as a game changer for people' lives, for the economy and all industries and services. The challenge is whether humans, members of the Gigabit society, will keep the control over the unbelievable and fast developing technologies or lose it with all still unclear and unpredictable consequences. How to tackle the ethical problems related to the AI activities? How to support humans and develop their openness (to avoid the natural lack of the openness) for interactions with AI?

Citizens' view and a certain level of the oversight of the AI development by all involved stakeholders is necessary for making sure that the new scientific studies (based on AI) will be developed in **agreement with humans needs**.

Fourthly, considerations about the economics of science development lead us to the challenge of the **excellence**.

The best scientific results will be a lost opportunity without mobilising all resources to build the "ecosystem of excellence" along the **entire value chain,**[5] starting from the research ideas and lab works, via creating the right incentives to accelerate the adoption of the solutions by all kinds of possible industries. Currently, and in the context of the AI development discussion, the "ecosystem of excellence"

---

[5]European Commission, White Paper. On Artificial Intelligence—A European approach to excellence and trust (COM(2020) 65 final), February 2020, part 1 Introduction.

ought to cooperate with the "ecosystem of trust".[6] This creates a special added val coming from the human factor of the science development.

What is the specificity of the **"ecosystem of excellence"**?

In recent years, the European Union debated heavily on the innovation policy mix needed. It was clear that the research and development policies should be strongly linked with industrial and SME policies, education, skills and regional policies (with the concept of the smart specialisation). At the same time, some frameworks were necessary to adjust the financial support, state aid, tax policies and public procurement models. Additionally, it became obvious that the most adequate regulatory framework should function more effectively: single market and competition rules, all types of regulations: strong/strict and soft, standards and the schemes for certifications, and last but not least—the intellectual property rights and the transparent system of patents. In many discussions, the soft dimension of EU innovation policy mix was emphasised: partnerships and coordination initiatives, and the culture of innovation.[7]

Some of those features are the key components of the supply side, some of them are crucial for the demand. This kind of view and understanding is significant for the economics of science providing for the **market conditions** for the science development.

Moreover, the "ecosystem of excellence" requires instruments, which could allow us to measure the effectiveness of the excellence in comparison to other science areas (the European Research Area in comparison to US, China, Japan, South Korea).

The following instruments can be considered: share of Science and Engineering publications in the top 1 citation percentile, general share of publications in global top 1% of cited publications, scientific excellence by fields, universities in Top 100 rankings, international mobility of researchers, shares of PhDs awarded and doctoral students in the US coming from EU, R&D spending, patent activity indexes (especially for selected technologies in selected economies), patent applications for domestically owned inventions made abroad, successful international collaborations, etc.[8]

When looking at these instruments, the general conclusion is unfortunately full of doubts and concerns, which puts the objective of making Europe the Global Centre for Excellent Research under constraints.

If Europe wants to **compete with leading powers at the global level,** a proper scale of the science development is indispensable. It requires overcoming the current fragmented landscape of centres of competences for the research development and full support from the EU level and the Member States to the European Research

---

[6]Ibidem, part 4, An ecosystem of excellence.

[7]EU Innovation Policy—Part I, Building the EU innovation policy mix, European Parliamentary Research Service, In-depth analysis, p.7.

[8]indicated and mentioned in the: Europe—the Global Centre for Excellent Research, Study requested by the ITRE Committee of the European Parliament, Reinhilde Veugelers, Michael Baltensperger Policy Department for Economic, Scientific and Quality of Life Policies, April 2019.

Area. It should be the most expected field and pattern of cooperation, taking all potentials from the scale into account. More synergies and networks between multiple European centres are necessary.

It must be established and supported by the European funding but it is critical how this funding is accessible. The size and scope of the funding should take into account the visible differences between some areas of Europe: the Eastern European countries are at the lower positions in all rankings, also with regard to the participation in the distribution of Horizon 2020 funded projects.[9]

It is clear, that the perspective of the economics of science could open not only the eyes of the policy makers in all European countries but also to wake the innovation potential of research centres in different countries (also lagging behind in terms of the science development) to pursue common collaborative partnerships. It should be done with some incentives, but also by showing the concrete advantages of building the common consortia. The concept of the common consortia means the participation of not only representatives from EU universities but also the representatives of the business from various parts of the EU.

The science/business partnership is a key condition for the commercialisation of the research works and is essential in the discussion on the economics of science. Additionally, it is clear that the above-mentioned consortia should have a business-oriented leadership. There is yet another, complementary reason to ensure the business oriented management: when analysing the potential of the European Science, we see—by looking at the number of citable documents[10] by subjects—a competitive advantage of the EU Science in nearly all areas. The activities of the scientists and the results of the researches are excellent, but the commercial use of them is low and remains insufficient. This leads us to safely conclude that despite the excellent scientific potential we are not using it at the needed European scale.

Clearly, this observation also shows that excellence should be visible at all stages of the value chain of the economics of science—from initial ideas, via lab works, through pilot projects to the commercialised implementation.

The tipping point for the European economics of science is therefore clear—**using all assets**: from the mature model of the **data infrastructure**, via full adoption of the **European Open Science Cloud**, openness for **citizens' participation** in building the knowledge society with human factor (addressing the new technologies development), and the establishment of the **"ecosystem of excellence"**.

It will lead us to the future: to the NEXT GENERATION of EUROPE, recovered and resilient.

---

[9]Ibidem, p.72 and 73.
[10]Ibidem, p.66.

The SKA Approach to Sustainable Research

Simon T. Berry

Contents

1 Introduction

The Square Kilometre Array (SKA) is an ambitious project to build a research infrastructure that will enable breakthrough science and discoveries not otherwise possible with current or planned astronomy facilities. Comprising two radio telescopes it will ultimately be the largest scientific instrument on Earth, both in physical scale and in terms of the volume of data it will generate. Like all infrastructures, there is a requirement for SKA to demonstrate where benefits have been realised from past investment, and the potential for future ongoing benefits from technology and knowledge transfer and innovations. The essay discusses the range of non-science benefits for the funders and for wider society going beyond SKA's core scientific mission and the challenges of thinking about regional impacts when designing a global research infrastructure.

SKA will have a uniquely distributed character: one observatory, operating two telescopes, on three continents for the global scientific community. The SKA's operational phase is expected to last at least 50 years.

S. T. Berry (✉)
Director of Corporate Strategy, SKA Organisation, Macclesfield, United Kingdom
e-mail: S.Berry@skatelescope.org

© The Author(s) 2021
H. P. Beck, P. Charitos (eds.), *The Economics of Big Science*, Science Policy Reports, https://doi.org/10.1007/978-3-030-52391-6_4

25

The recently completed SKA Global Headquarters is located at Jodrell Bank near Manchester in the UK, home to the organisation—the future SKA Observatory—that oversees development, construction and operations. The two other SKA sites are radio quiet zones and home to the telescopes themselves: a mid-frequency array in South Africa (SKA-mid), and a low-frequency array in Australia (SKA-low). In order to take advantage of the development of computing and other innovative technologies of relevance for the SKA programme, the construction of the SKA will be phased. Work is currently focused on the first phase named SKA1.

SKA1 will be a transformational scientific facility. It will tackle some of the most fundamental scientific questions of our time, ranging from the birth of the Universe to the origins of life.

To do so, the SKA will collect unprecedented amounts of data, requiring super-computers amongst the world's fastest to process this data in near real time, before turning these into science products for distribution around the world through a network of SKA Regional Centres located in partner countries. Those data centres will be the final interface with the end users—the scientists—who will turn these science products into information, and finally knowledge. The two SKA telescopes differ in design and are complementary by their very nature. Both are interferometers: arrays of antennas which when linked together act as one enormous telescope, bigger than would ever be possible in a traditional single-dish design.

In South Africa, the Design Baseline for SKA1-mid comprises 197 dishes, 64 of which are already in place and form the MeerKAT precursor telescope, itself a world-class facility, which will be integrated into SKA1-mid. In Western Australia, over 130,000 low-frequency antennas will form SKA1-low, spread across 512 antenna stations.

Infographics describing SKA1-mid and SKA1-low telescopes

The telescopes' design is scalable and upgradable, allowing future improvements to maintain their world-leading capabilities, and also to align with available funding. This includes state-of-the art scientific and computing infrastructures, designed to progressively exploit the capabilities of the Observatory as computing technology continuously improves over the coming decades.

2 SKA Science

The SKA's science goals are broad and ambitious, looking back into the history of the Universe as far as the Cosmic Dawn, when the very first stars and galaxies formed, in order to seek answers to some of the biggest remaining questions in astrophysics. Among them: How do galaxies evolve? What is dark energy and what role does it play in the expansion of the Universe? What causes planets to form around stars? Can we find and understand where gravitational waves come from? Is there life out there? Individually, and working together with other next-generation facilities, SKA will deliver a profound impact.

The SKA will devote between 50 and 75% of its time to Key Science Projects— major surveys that require lots of observing time on the telescope, focused on making ground-breaking discoveries. The remainder will be allocated to traditional, smaller-scale studies depending on scientific and technical feasibility. Access to the telescopes will be on merit, but with the majority of time being reserved for scientists from the Observatory's member countries.

As with any major scientific endeavour, the SKA's design has undergone changes during the course of the project to take account of technological evolution and funding constraints. In July 2013, prior to the start of detailed design, the SKA Organisation Board set a cost cap of €650 m for the construction of SKA1. This has been periodically adjusted for inflation and, based on the most recent adjustment (December 2017), is set at €691 m, representing the 'Deployment Baseline' of the infrastructure. As of June 2019, the cost estimate for SKA1's Design Baseline, which is the agreed design for the first phase of the SKA project described above, is €914 m (2017 €). The cost estimate is under regular review and will be subject to a full cost audit in conjunction with the Critical Design Review of the overall SKA1 system. Over the coming months, alongside the implementation of the organisation to deliver SKA, discussions will focus on available resources against the project cost to establish a final project scope.

A high-level view of the technical and scientific milestones for SKA Phase 1 is captured in the figure below.

3 Sustainability

SKA has worked intensively to align itself with the outputs of the ESFRI working group on RI Sustainability. Sustainability for the SKA means a combination of financial stability, a governance environment that will endure through the project's lifetime, and a series of measures that ensures SKA sits sustainably in its host countries and delivers a compelling socio-economic benefits case for its membership.

4 Governance

Following a familiar route for European Research Infrastructures, SKA has evolved from an informal collaboration through to the establishment of a (UK) company to manage the design effort preparing for construction. Then, following a decision taken by the SKA member governments in 2015, work began to develop an intergovernmental organisation (the SKA Observatory) to manage the construction and operation of the facility. Enabling a long term, sustainable governance structure has been at the heart of the decision.

The IGO will operate in a similar way to other long-established intergovernmental infrastructure organisations such as CERN, ESA and ESO and is being designed taking into account best practice from all relevant European and global examples. Such organisations embody a commitment by the Members to the long-term financial and operational arrangements that are essential for the secure delivery of large complex research programmes and facilities. It will move from its present status to that of a treaty-level organisation registered with the United Nations in early 2020, and will be governed by a Council appointed by the Members and acting collectively to advance the project in accordance with the principles set out in the Convention that will be its founding document. It will be established and will operate under international law, with the appropriate legal status and privileges in the Member countries.

5 Socio-economic Benefits from the SKA

The creation of large scientific facilities, while driven by clear and compelling scientific needs, leads not only to scientific advances but also to a range of non-science benefits for the funders and for wider society. The 2015 report by "Science Business" entitled "Big Science: What's it Worth?" identified a range of benefits arising from investment in "Big Science", from capacitive touch screens to cochlear implants, and also presented a set of case studies looking in more depth at the impact of Big Science projects (including SKA) beyond their brief. A variety of other studies have also addressed the issue of non-science benefits arising from the major investment entailed in such facilities. Like all infrastructures, there is pressure on SKA to demonstrate where benefits have been realised from past investment, and the potential for future ongoing benefits from technology and knowledge transfer and innovations.

From the earliest ideas, and throughout the development of the SKA concept, close attention has been paid to the potential non-science benefits of the Project. The overall aim is to ensure that there is a progressive and persistent case for investment that goes beyond the science case, based on innovation and industrial return. The evolving design and construction plans endeavour to facilitate identification, exploitation and propagation of innovation and other non-science benefits. Moreover, SKA has a potentially unique opportunity to establish a structure for identifying and tracking all relevant metrics for impact from outset to conclusion.

The first and most immediate benefits from investment in SKA are guaranteed access to direct contracts and the opportunity to make in-kind contributions, both of which are only open to Members of the SKA Observatory. Additionally, the very fact of working with such an advanced project can lead to companies developing existing and new skills that enable them to be more competitive. Long-established evidence from companies working with other cutting-edge facilities (e.g. CERN*) has shown that responding to the exacting standards demanded has raised their game and qualified them to bid for other high-technology contracts, thus extending their markets.

6 Socio-economic Benefits

The demands of the SKA will help drive technology development, and contribute to and be a stimulant for invention and innovation within global industries. In conjunction with market forces, they will lead to benefits to society that will be widely applicable in other established fields and will begin even before the astronomical discoveries that will eventually be made using the telescope. Countries that participate in the SKA could reap considerable socio-economic benefits by participating at the forefront of these developments. Spin-off innovations in areas specific to the SKA's computing activities will, through the industries with which SKA will be working to develop them, benefit other systems that process large volumes of data from geographically dispersed sources. Potential areas where innovations inspired by the SKA's needs could have wider applicability include data management techniques, data mining and analytics, imaging algorithms, remote visualisation and pattern matching (all of which will have impact in areas such as medicine, transport and security).

The significant contribution that the SKA will make to new developments across a range of technologies will also benefit the global knowledge-based economy beyond scientific research and academia. For the first time, the developing world is an active and integral contributor to fundamental research on an unprecedented scale. Emerging and developing countries are already active in the production and exchange of knowledge, stimulating their participation in the global knowledge-based economy. In addition, the existence of the facility will create demand leading to employment opportunities such as visitors' facilities in surrounding towns to cope with the expected interest from astro-tourists; science centres; development of curriculum material related to the SKA; the inspiration of popular culture, etc. Further afield, local opportunities will arise around the users' home institutions in response to the work they will be undertaking remotely with the telescopes, ensuring that SKA has a truly global presence. The potential reach of the SKA in terms of skills development is therefore enormous.

The SKA will need thousands of people from across many countries in a range of professions to design, build and operate it and to run all the associated services. Builders, caterers and administrative staff as well as scientists, engineers, IT specialists, communicators, etc. are already being employed in this endeavour, encouraging the creation or enhancement of a broad range of training and skills development opportunities. Of particular interest both for the immediate and specific needs of the SKA and for the future extension of knowledge-based economies is the development of STEM skillsets.

Sustainability is also about skilled people, both supporting the project's implementation, and arising from the project's implementation. To ensure a sufficiently large pool of trained scientists, engineers and technicians by the time the SKA is ready to be operated, it is urgent to inspire young people now. By working closely with education authorities as early as possible, building on the experience and success of existing efforts, in particular NASA/HST or astronomy education programmes such as Universe Awareness (UNAWE), the SKA can make a substantial contribution to current efforts to bring astronomy onto the school curriculum in a number of countries. The SKA Organisation and its Members are already actively engaged in developing programmes in pursuit of these aims. The mounting concerns about the widespread shortage of trained scientists, engineers and technicians are not limited to the SKA and its own needs. These skills are critical to countries' increasingly technologically dependent economies, and governments are keen to find ways to promote them. The active engagement of the SKA Organisation/SKA Observatory IGO in generating wider interest in and support for training in these subjects, including strengthening recognition of the need for and value of them even among those who will not directly follow this path, will contribute to wider economic advancement as well as meeting its own specific needs. The SKA can become the first international scientific infrastructure with education and outreach embedded in its development from the earliest stages, inspiring young people in time for them to become users of the telescope or engineers and scientists working with the SKA, and potentially encouraging more people to develop he STEM skills needed to maintain and grow the knowledge-based economy.

7   Case Study: SKA South Africa HCD Programme

In 2005, South Africa established a Human Capital Development (HCD) Programme to create capacity in relevant radio astronomy science and engineering. The programme, which is available to South Africans, and to students from SKA SA Partner Countries in Africa, provides support at all academic levels, to ensure a continuous throughput of young people moving into relevant studies and research, and to ensure the required supervisory and teaching capacity is in place to support the students. To date, the programme has awarded more than several hundred grants to students and universities. In addition to the research capacity development aspects of the programme, SKA SA has also focused on developing skills for the operations and maintenance of the MeerKAT facility. The Artisan and Technician Training programmes have well defined schedules of the numbers of technically skilled staff required, and are on target to meet the capacity needs. In the towns that are close to the SKA SA site in the Karoo, the SKA SA HCD programme has facilitated the recruitment and secondment of Mathematics and Science teachers in the schools, provided bursaries for learners to attend Carnarvon High School (in the town nearest the site), rolled out cyberlabs and E-learning centres (in cooperation with the Departments of Education and Rural Development, and with industry partners), and trained local residents to manage cyber-centres, which are available to members of the community.

8 Conclusion

Sustainability for a research infrastructure means being able to motivate a continued significant investment to support activities. It requires a multidisciplinary approach which brings together a suitable governing legal framework, a policy environment which permits financial stability, all underpinned by a scientific and socio-economic case. SKA is working hard in all domains, with many aspects to come together with the formation of the SKA Observatory IGO and approval of the telescope's construction in 2020.

Full Presentation

https://indico.cern.ch/event/727555/contributions/3461263/attachments/1868120/3072746/FCCW_0900_Berry_SKA_approach_to_sustainable_research.pdf

The European Spallation Source: Designing a Sustainable Research Infrastructure for Europe

John Womersley

Contents

1 Introduction

The paper briefly outlines some of the key challenges in building sustainable support for any science megaproject, using the European Spallation Source (ESS) as an example (Fig. 1). Beyond the project's imminent socio-economic impact the essay also reflects on the broader question of how public investments in large-scale "Big Science" projects can tackle the present global inequalities by reshaping the current forces of globalization, offering more opportunities for participation and empowering marginalized groups that often feel excluded.

Any science megaproject needs to have the following in place to succeed:

- a solid science case
- technical R&D carried out to reduce key risks, and cost estimates that are understood
- a solid project management plan
- a credible funding and governance plan
- good stakeholder engagement and support, and
- a compelling investment case

Let's look at each of these questions in turn.

J. Womersley (✉)
The European Spallation Source, Lund, Sweden
e-mail: John.Womersley@ess.eu

H. P. Beck, P. Charitos (eds.), *The Economics of Big Science*, Science Policy Reports, https://doi.org/10.1007/978-3-030-52391-6_5

Fig. 1 ESS under construction in Lund, Sweden. The central concrete structure will house the neutron production target

2 Designing the European Spallation Source

For ESS, the investment case in a nutshell is that scientific and technological innovation is essential to address both the *global challenges* of energy, climate, environment, healthcare, and the *economic and societal challenges* of stalled productivity and long term wage stagnation. The particular contribution that ESS will make is to use intense beams of neutrons to allow scientists to study the structure of materials and molecules: where are the atoms and what do they do? This knowledge underpins the development of new materials, new drugs, new processes, and new energy technologies.

The vision of ESS then is to build and operate the world's most powerful neutron source, enabling scientific breakthroughs in research related to materials, energy, health and the environment, and addressing some of the most important societal challenges of our time.

There is a long and successful legacy of neutron scattering worldwide, based primarily on the use of research reactors as neutron sources. However, there is an effective maximum flux of neutrons that can be obtained from a reactor if the chain reaction is to be controllable. The way to generate much higher fluxes, to study smaller samples or make measurements faster, is to use an accelerator-driven source that produces neutrons through spallation reactions from a high-Z target (tungsten in the case of ESS) when struck with a. proton beam of energy 1–2 GeV/c^2. The spallation neutrons emerge with high energies characteristic of nuclear processes and therefore need to be slowed down until their quantum mechanical wavelengths are of the order of 10^{-10} m, the spacing between atoms in the kind of samples that we wish to study. This is done by elastically scattering the neutrons; ESS uses a polarised liquid hydrogen moderator to achieve this. The neutrons are then guided to experimental stations where the samples to be studied are located, and the scattered neutrons are detected and measured by large instrumentation arrays. From the angles, energies and phases of the outgoing neutrons, the structure and composition of the sample can be inferred.

Some areas where technical advances and R&D were needed to deliver the ESS performance goals were:

- the superconducting accelerator and in particular the use of spoke cavities in the low-energy section
- the neutron production target which is based on a helium cooled rotating wheel
- the neutron moderator design which is two to three times more efficient in terms of useful neutrons per unit beam power and
- the development of solid state boron-based neutron detectors to replace increasingly scarce helium-3.

All of these innovations required early-stage R&D funding to allow their development to the point where the risk of proceeding with the full project was acceptable.

A few numbers set the scale of the project. The construction cost (in 2013 prices) is 1843 M€. At 5 MW (2 MW at start of operation) ESS will operate the world's most powerful particle accelerator—a 400 m long superconducting linac—to drive the spallation target. Neutrons will be detected in 15 experimental stations, each optimised for different types of measurements. The combination of increased accelerator power, more efficient neutron generation and better resolution instrumentation will yield an increase in sensitivity of a factor of 20 on average compared with today's best facilities of this type. We anticipate roughly 800 experiments will be carried out per year, by several thousand visiting scientists.

Groundbreaking took place in Lund, in southern Sweden, in 2014, and the project is now 60% complete. 2019 is the peak year of construction activity. Major installation activities are underway and we have entered the initial operations phase, with the first components of the particle accelerator in commissioning. First science with external users will take place in 2023.

To establish a new facility like this on a green-field site requires not just constructing the scientific infrastructure but also the organisation to build and operate it. ESS has recruited over 450 staff from 54 nationalities to work on the project, roughly 40% from Sweden, 40% from the rest of Europe and 20% from the rest of the world, including from as far away as Japan, Australia, the USA and Canada.

The Legal structure of ESS is a European Research Infrastructure Consortium (ERIC) which was conceived as a lighter weight alternative to setting up a treaty organisation for European inter-governmental scientific facilities. The ESS ERIC was established in 2015. There are 13 member states. The two host Countries, Sweden and Denmark, have agreed to pay a 47.5% share of the construction costs, mainly in cash; the non-hosts pay the remainder but roughly 70% in kind rather than in cash.

Fig. 2 One of the first major in-kind deliveries at ESS. The proton source, built at INFN in Catania in Italy, is officially inaugurated in Lund by HE the President of Italy and HM the King of Sweden

By in-kind contributions we mean significant technical workpackages procured or constructed in partner countries rather than centrally. The partner holds and manages the risk. This can be a project management challenge, but nowadays is close to a political necessity. It helps avoid the situation where the host region benefits greatly and the others just pay cash. At ESS, 70% in kind for the non-hosts, translates to about 35% overall, which is just about manageable as there are sufficient cash contributions to cover the needs of project contingency. In contrast, the ITER Project, where the in-kind fraction approaches 90%, has proved very hard to manage (Fig. 2).

ESS has established a network of in-kind partners in each of its 13 members. Doing so has benefited the project by providing access to well-established technical expertise and engineering resources in national laboratories such as Saclay, RAL, Jülich and avoided the need to establish any manufacturing capability for accelerator and target components in Lund. Managing the in-kind supply chain, though, including procurements, quality control, schedule and eventual systems integration at ESS, does require significant effort.

Any large capital investment needs to show that it is following good project management practice. In ESS's case we have chosen to mirror the requirements of the US Department of Energy's Office of Project Management Oversight and Assessment (DOE order 413.3B) which is usually seen as best practice in the big science domain. Any project must have a resource-loaded schedule, sufficient contingency, a clear change control process, and good financial controls. ESS uses Primavera as our project planning tool and the ESS schedule contains over 20,000 linked activities. The critical path is clearly understood and proactively managed.

3   Meeting the Modern Challenges

Up to now we've discussed requirements that would have applied to any major science investment over the last few decades. But a new facility being built in 2019 must also reflect today's expectations. It must meet modern environmental expectations: ESS is a green construction site with no waste to landfill and construction machinery powered by biofuels. We will purchase all our electrical power from renewable sources, and waste heat from our electrical machinery will be recovered into the Lund district heating system. Modern research facilities must also meet modern data handling expectations.

ESS will provide support for the full computing, software and analysis chain at its data centre in Copenhagen and will operate an open data model as part of the European Open Science Cloud. Modern facilities are also expected to help the translation of research from academia to industry; to this end a science and innovation campus is being developed in the land area between ESS and the neighbouring MAX IV light source.

We have now briefly reviewed most of the items listed in the checklist described at the start—having a good science case, technical R&D and cost estimates understood, project management, a credible funding and governance plan, and a compelling investment case.

4   Concluding Remarks

Finally, a few closing observations. Firstly, why do big projects fail to get started? In 2016 the European Strategy Forum for Research Infrastructures (which I chaired at that time) reviewed the implementation progress of the projects on its roadmap. We found that **inadequate stakeholder engagement** and **lack of a credible funding plan** were the biggest barriers to implementation—much more so than any weakness of the science case.

Stakeholders include (but are not limited to) the General Public, media and "opinion formers," students, educators (the STEM skills pipeline),university bosses, other areas of science, Members of Parliament, Science Ministers, Finance Ministers, Opposition political parties, Local and regional politicians, Civil Servants, economists. All of these are multiplied by the number of member countries, each with their own science strengths, industrial profile, media and decision-making culture.

Especially where the government stakeholders are concerned, there has been a shift in emphasis since the end of the Cold War from investing in science to promote normative values (implicitly in contrast to the values of the competing bloc) such as openness, the cultural value of science, science for peace and democracy, and to promote international collaboration, towards market or social values such as science as a driver of prosperity, jobs, innovation, startups and as a solution to grand challenges for climate and the environment. Scientists were probably rather happier with the cold war values, as they made fewer demands on them—but investment decisions are now made firmly on the market and social impact of projects.

The biggest economic challenges of our time include globalisation, which together with automation and new technologies is leading to fewer good jobs in developed countries, to low growth, stagnant wages, angry people on the streets and populist politics. No one really has the answer to this, but there is a general consensus that scientific innovation and STEM skills are key—or at least economies and people that have these skills will be better positioned to deal with these challenges. So what is our project going to do to help? This is a question that all of us as scientists, but also policy-makers, citizens and especially students who will be the future users of these facilities need to be able to confidently answer.

Further Reading

1. European Spallation Source – 2018 Activity Report: https://europeanspallationsource.se/sites/default/files/files/document/2019-08/ESS%20Activity%20Report%202018.pdf
2. European Strategy Forum on Research Infrastructures – Roadmap 2016 https://www.esfri.eu/sites/default/files/20160308_ROADMAP_single_page_LIGHT.pdf
3. Neutron users in Europe: Facility-based insights and scientific trends: https://europeanspallationsource.se/publications

Full Presentation

https://indico.cern.ch/event/727555/contributions/3476197/attachments/1868049/3072618/JohnWomersley.pdf

Optimising the Benefits from Research Institutes

Thierry Lagrange

Contents

Today, a large number of public research institutes have been set up in many different fields to carry out scientific research. The initial financial investment needed was generally justified by the pure scientific interest. As the number and size of these research institutes has grown over time, the associated investments have become quite substantial. The will to push the frontier of knowledge implies developing and upgrading high-tech instruments, working with cutting-edge technologies, and high maintenance costs, which requires additional public spending. As the available public resources have not evolved at the same rate as scientific ambitions, competition between scientific fields and projects has become fiercer as well as the need for additional arguments to justify these investments.

The pressure to show the potential social benefits beyond the scientific case have therefore intensified, making it increasingly difficult for projects to get funding or remain financially sustainable without socio-economic justifications.

Each research field is different and the socio-economic benefits can vary widely from one field to another. The nature of the research, whether it is fundamental or applied research, whether the instruments used require vast financial investment or not, and whether breakthrough technology is required or not will result in different socio-economic benefits.

T. Lagrange (✉)
Procurement & Knowledge Transfer Department, CERN, Geneva, Switzerland
e-mail: Thierry.Lagrange@cern.ch

© The Author(s) 2021
H. P. Beck, P. Charitos (eds.), *The Economics of Big Science*, Science Policy Reports, https://doi.org/10.1007/978-3-030-52391-6_6

1 The Case of CERN

CERN exists to carry out fundamental research in high energy physics. Its activities necessitate large investments, breakthrough technologies, periodic upgrades and maintenance. The socio-economic benefits are multiple: training of scientists and engineers, outreach and education, development of innovative technologies and collaboration with industry, to name just a few. The scientific justification and the fact that fundamental science is indispensable for applied research is not disputed, but funding institutes expect that all other possible benefits be optimised too.

The challenge, therefore, is understanding how to accelerate and optimise the generation of these benefits. It is clear that time is of essence: the sooner society can benefit from spin-offs generated by research institutes, the bigger the impact will be on further research funding and on global growth. At the same time, most of the technologies developed in connection with CERN's equipment and operations have too low a technology readiness level (TRL) to trigger the immediate interest of private capital or big institutional investors. For this reason, the process of bringing CERN's innovation to society is seldom spontaneous and direct. In order to be effective and systematic, the research institute itself needs to actively engineer improvements in the TRL, supported by public funding, until it is sufficient to attract external capital.

2 Building an Ecosystem

Building a proper ecosystem is an indispensable prerequisite to transferring knowledge and technologies to society in a way that maximizes the socio-economic benefits. Such an ecosystem thrives with the generation of innovative, preferably breakthrough, technologies, resources to increase the TRL, a fit-for-purpose IP policy and governance, access to a network of technology brokers, incubators, venture capitalist and policy makers, and the active support of the scientist or engineer at the origin of the innovation. The involvement of knowledge transfer experts is also essential. The CERN Knowledge Transfer group provides advice, support, training, networks and infrastructure to ease the transfer of CERN's know-how to industry and to society.

Collaborations and Networks
Knowledge transfer networks
Strengthening links with Member States (KT Forum)
Relations with International Organisations
Knowledge transfer in EC co-funded projects

Entrepreneurship
Start-ups & Spin-offs
Entrepreneurship Meet-Ups
Business Incubation Centres
Entrepreneurship Programmes

Events
Knowledge Transfer Seminars
Conferences with a significant contribution by the Knowledge Transfer group

Intellectual Property Management
R&D collaborations
Patent portfolio
Licence, service & consultancy agreements

Funding Opportunities for CERN Projects
CERN Knowledge Transfer Fund
CERN Medical Applications Budget

Open Source
Open Source Software
Open Hardware Licence

Support for CERN Personnel
Formal and practical training in business, entrepreneurship & knowledge transfer
Legal, business & intellectual property support

CERN Knowledge Transfer Ecosystem

Similar to other institutes, CERN has created a business incubator network. This involves CERN signing agreements with existing incubators in its Member States to facilitate the creation of start-ups based on CERN technologies. The incubators promote CERN's technologies and give the start-ups access to their full range of support services. In return, CERN makes available to the start-up the necessary IP and initial technical support, plus the right to use CERN's branding.

It should be noted that for scientists and engineers behind the technologies developed at CERN, identifying industries or sectors where those technologies could be best applied is far from straightforward. For this reason, CERN has increased its contacts with early-stage venture capital firms. These firms bring a complementary perspective: they typically start from 'real life' so-called 'pains' in domains such as health and transport, determine the limitations of the existing technologies, and then search for technologies that can bring solutions to well-identified problems.

Large companies also face similar problems and some are looking for solutions (or for help to find solutions) in academia. A key factor for successful technology transfer is identifying an industry problem that can be addressed by the technologies and competences of research institutions (market pull), as opposed to simply pushing technologies which have been developed to solve specific research-related problems (technology push). Finding this match between industry and research is often a recursive process. Discussions often take place with several companies before a good match is found with some of them.

3 Technology Transfer Through Procurement

When discussing the socio-economic impact of big science centres one cannot ignore the role of procurement in constructing, operating and maintaining the sophisticated scientific tools but also for meeting the needs of thousands of researchers from around the world that use CERN's infrastructure for their research.

CERN has an average annual expenditure on goods and services of 500 million Swiss francs. Among other things, CERN procures civil, mechanical and electrical engineering, information technology, radiofrequency, cryogenics and vacuum as well as detector technologies. When the Organization needs to buy innovative equipment, it has the option of either doing the development itself and subcontracting the series manufacturing, or entrusting industry with the whole process from design to production.

In order to reduce the cost and risk associated with high-tech innovative equipment for which CERN has established know-how and expertise, the development phase is often done in-house. The contract for the series production is then based on a build-to-print specification written by CERN. It is at this stage that the technology transfer happens from CERN to the industry. On its side, industry brings the expertise and know-how needed to scale a prototype operation into a series production.

For firms, this technology transfer can result in the development of new products, improved products or services, better technical knowledge in their field and the ability to reach new markets. After working with CERN, companies often invest more in R&D, file more patents and exhibit higher levels of productivity.

4 Conclusion

Literature and studies confirm the high socio-economical potential benefits from research institutes to society. These benefits are not generated automatically, they require a process that includes the selection of the most promising technologies to the support to bring to maturity the technology to the market. This support can take many different shapes according to the needs. Which IP, which resources, what kind of network, which skills are required, this varies case by case, but to be successful, all of them must be covered.

Further Reading

1. CERN KT's 2019 Annual Report: https://kt.cern/about-us/annual-report
2. Impact of CERN's Procurement Actions on Industry: 28 illustrative stories (CERN, 2019): https://cds.cern.ch/record/2670056
3. CERN's 2018 Annual Report: https://cds.cern.ch/record/2671714?ln=en

Rethinking the Socio-economic Value of Big Science: Lessons from the FCC Study

Johannes Gutleber

Contents

1 Introduction

Investing in fundamental research is often considered a risky venture. The associated costs for designing, developing and building new scientific instruments, the long timelines for the construction and operation of these facilities and the sophistication of the enabling technologies—often calling for further R&D investments to meet the market needs—are among the fear-factors that enter into the debate around the investment in fundamental research.

This is also the case for High-Energy Physics. To study nature with higher precision and to understand the fundamental building blocks of our Universe, we need high-performance particle colliders. The construction and operation of the LHC serves as an example. Yet it often goes unnoticed that these large-scale research instruments can also offer positive returns for economy and society as well as many opportunities for industry and enable co-innovation through international collaboration among academic centres and laboratories. We note a similar picture in other areas of fundamental research like astrophysics or the emerging field of gravitational waves astronomy that also call for large-scale research infrastructures and significant public investments.

In the course of the twentieth century, we have witnessed how big research facilities generate a focal point for collaboration among a multitude of actors from

J. Gutleber (✉)
Directorate for Accelerators & Technology, CERN, Geneva, Switzerland
e-mail: Johannes.gutleber@cern.ch

© The Author(s) 2021
H. P. Beck, P. Charitos (eds.), *The Economics of Big Science*, Science Policy
Reports, https://doi.org/10.1007/978-3-030-52391-6_7

academia and industry. Benefits stemming from large-scale facilities, often go beyond the pure scientific knowledge we gain from them about nature. They act as hubs of innovation and technological and scientific collaboration, alongside their core scientific missions. They regularly enable synergies among the producers of knowledge and the industrial partners or innovators that turn out close to market products. History has shown that big scientific facilities consistently yield surprises that in turn are converted into products while sometimes giving birth to whole new industry sectors.

Science and technology underpin much of the advance of human welfare and the long-term progress of our civilization. This is reflected in an extraordinary growth in public investment in science. At the same time we see a rising demand to demonstrate the societal return from these investments. The rising interest on the type of benefits that emerge and how they can be maximized and redistributed to society gives rise to a new field of interdisciplinary research bringing together economists, social scientists, historians and philosophers of science and policy makers.

2   Large-Scale Research Infrastructures for Particle Physics and Beyond

Let me focus on the field that I have been working for the past decades, namely accelerator-based research infrastructures. Today, we are at a critical moment for fundamental physics following the discovery of the Higgs and the first observation of gravitational waves, both opening new windows in our quest to understand the Universe. At the same time we are entering a critical stage for shaping the science policies that can tackle the scientific challenges of the twenty-first century. In the European landscape we witness the ongoing discussions about an ambitious post-H2020 framework programme for Research & Innovation, the establishment of the European Innovation Council and the adoption of a mission-oriented policy to bridge the gap between the research cloud and market needs to name but a few of the ongoing processes. We also observe similar debates and transitions taking place at a global scale with new countries raising their public investments in big research infrastructures.

In fact, there is a rising consensus that research infrastructures have a broader return for society beyond their core scientific mission. This is why in my view we should not give up on quantifying and furthermore strive to maximize the socio-economic impact generated from Big Science facilities.

This point bring me to the second question that I would like to touch during this short presentation; namely how much basic research is needed to go beyond 'new fundamental knowledge' and achieve what one calls 'usefulness'? Can we afford as a society a continuous investment in curiosity-driven research that doesn't promise precise and immediate applicable results? These are questions that more and more inform the public debate and remind us of the political and societal importance of science while calling for continuous efforts to communicate and disseminate our results. As we are moving towards a new science-oriented economy of knowledge and innovation we need to revisit the role of RI's and the ecosystem that they support. This means that we need tools and more data that will allow design RIs in a way that could maximize their socio-economic impact and policy-makers to arrive at more informed decisions.

Designing a global-scale project, like the Future Circular Collider for a post-LHC collider-based infrastructure unavoidably triggers discussion on how we can quantify the benefits stemming from this project for the involved industry and academic partners, beyond the key scientific questions that we can explore with this facility. In the past years, together with collaborators from other universities and research institutes we have tried to identify the key returns for society from public investments in large-scale projects like LHC and its high-luminosity upgrade (HL-LHC) and use these results to forecast the impact of the planned FCCs. These results have been published in a number of papers and I will briefly discuss them below. But let me add, that this exercise also revealed the need for new thinking from scientists, economists and policy makers which is what motivated us to organize this workshop.

3 Measuring the Socio-Economic Impact: Challenges and Prospects

In the following few paragraphs, I would like to share some of the insights that we have gained in the past years through our work for the FCC study and the supporting H2020 projects EuroCirCol and EASITrain. The FCC study was launched in 2014 to prepare the ground for a post-LHC research infrastructure and push R&D lines for technologies that could guarantee a sustainable and cost-efficient construction and operation. After 5-years, the project succeeded in building an international collaboration with more than 150 institutes from around the globe and delivered a four-volume Conceptual Design Report[1] that describes in detail some of the topics that I will highlight here.

The socio-economic impact assessment of the LHC/HL-LHC programme, carried out in the scope of an European Investment Bank (EIB) project by the University of Milano (Italy), has revealed the added value of public investment in research infrastructures. This was the first application of this method and gave us some encouraging results to reflect on how this impact can be better measured but also on the tools that would allow to further maximize it. Today, the H2020

---

[1]Future Circular Collider, Conceptual Design Report Volume 2, The European Physical Journal Special Topics, Volume 228, pages 755–1107(2019).

Future Circular Collider Conceptual Design Report Volume 3, The European Physical Journal Special Topics volume 228, pages 261–623(2019).

All the volumes of the FCC CDR can be accessed online here: https://cern.ch/fcc-cdr

EuroCirCol project is a reference case to apply the EU recommended framework for infrastructure CBA to the research community.

Findings of this analysis suggest that training is the single most important generator of socio-economic impact from such endeavours. Let me give an example that stresses the value of training in the different fields linked to the construction, operation and maintenance of an RI. The average salary premium of students involved in a large-scale particle accelerator research project across sectors and domains is between 5% and 13% in addition to the premium of obtaining a higher-education level academic degree (Master or doctoral degree). A conservative approach to translate this into an absolute monetary value yields a life-time (we estimated 40 years of professional active period) added premium of 150,000 Euros on average per student or early stage researcher.

In addition, this work identified the possible lack of a new generation of well-trained lead engineers in the domains of superconductivity and cryogenics. Maintaining a pool of skilled engineers and scientists in these fields is crucial not only for high-energy physics but for the more than 50,000 accelerators that operate worldwide from the pharmaceutical and food industry to global transportations and medical treatment. Consequently, the FCC project consortium submitted the EASI TRain (European Advanced Superconductivity and Innovation Training Network) proposal for a MSCA training network that was accepted in 2017. The project will continue running until June 2021, training a new generation of experts and leaders in the while establishing a curriculum that will serve as roadmap for future training & education in these fields. Moreover, many of the early stage researchers will pursue careers in other fields, outside particle physics, thanks to the skills that they will acquire during the EASITran and the network that they develop in the course of this project.

Beyond training, there is also a dominant effect from public investment in such facilities for the industry as also demonstrated by some of the previous speakers. Industries profit most via co-innovation and co-development in the mid- to hi-tech sector and for cutting-edge technologies that are brought to maturity. Big science centres like CERN are highly complex collections of instruments and installations, and invest heavily in the development of specifications of highly advanced technologies. They carry out large-scale construction projects according to strict plans and shared objectives, which are passed on to industrial suppliers in technical specifications as part of procurement procedures. Innovation and knowledge spillover between big science centres and industry is a process of interactive learning in a mutual relationship based on the complementary resources and objectives of the two organisations as previously shown[2].

Last but not least, an interesting finding from this work is the high value of cultural goods (science tourism, books, films, exhibitions) stemming from such large-scale facilities. Cultural effects, while uncertain because they depend on future announcements of discoveries and communication strategies, were estimated to

[2]Autio, E., Hameri, A.-P., & Vuola, O. (2004). A framework of industrial knowledge spillovers in big-science centers. Research Policy, 33(1), 107–126.

contribute 13% to the total HL-LHC benefits. More than half of this percentage comes from onsite visitors to CERN and its travelling exhibitions. Science is a global affair, but of course the love about science and the public appreciation remains bound to regional factors that should be further understood and explored. The cultural goods stemming from Big Science can also be a ground for synergies with other existing or planned RIs. Here one should also mention the willingness of taxpayers to support such a research infrastructure with comparable high amounts per year (order of 4 Euro/year/person) though the study revealed a strong correlation of the willingness to pay with the educational level of the citizens as well as their knowledge about the goals and scope of the planned infrastructure. This bring me to the last point of my talk about the value of communicating our work and engaging with the public throughout the lifetime of a project.

All in all, it is clear that understanding the socio-economic impact of Big Science demands a large-scale institutional response. There is a colourful landscape of impacts that come out from public investments in such facilities long before—and on top—the scientific lessons we gain. The applied methodologies and the interpretation of results should be a major subject in public policy, and at grant agencies and universities. I hope that this workshop and the discussions we have will contribute to this direction.

4 Future Outlook

Let me conclude this brief summary by pointing out a few possible avenues for further analysis that have been identified through our previous work in the framework of the HL-LHC project and the FCC study. As it is often the case in scientific inquiry, trying to answer one question generates more. However, I think that these points could help in shaping a common agenda for other RIs managers and policy makers and enable collaborations between different scientific projects:

- Analyse the wider social returns to non-R&D intangible investment. Evidence to date has focused largely on spillover returns from public and private R&D investments. This approach neglects the cultural and educational impact with the training aspect should be further explored and understood. This also calls for innovative approaches like the one presented by Prof. Loureiro and her colleagues using Big Data from social media.
- Further work to understand the wider 'public good' benefits of publicly-funded knowledge investments. Of course, the impacts of public R&D on a high-field magnet, a novel acceleration technology or a superconductor are extremely difficult to measure, but further attempts to do so would be useful and add significantly to the evidence base on the returns to knowledge investments. We also understand that it is hard to quantify the impact of inventions like the WWW but there are similar other developments where HEP is significantly involved

(i.e. detector technologies, computing, material science) and these contributions should be quantified.[3]

- Evidence on how firms in non-R&D intensive industries innovate and draw on public investments to do so through joint activities with Research Infrastructures.[4,5] The exposure to an international environment, the impact on a company's brand name are among the additional impacts coming through joint activities with a big research infrastructure.[6]
- The regional and global impact of the existence of a big Research Infrastructure in non-R&D intensive industries. Sectors like retail, financial services, transport and utilities represent large parts of the economy, and profit in different ways from the existence of a research facility in a specific area.
- Clearer evidence on the impacts of interaction between different forms of knowledge investments, not least the impact on returns, would be helpful.[7] It is never easy to isolate linear cause-effect 'returns' on research investment, but more granular evidence on these synergies and complementarities can always help.. This should include a combination of detailed data analysis of how R&D investments on new technologies spread across the whole value chain of a product and/or among different industries, and case study evidence.
- Given the emergence of different approaches for assessing the societal impact of RIs in different countries, and that by nature knowledge investments are highly internationally mobile, it would seem important that there is an attempt to coordinate across countries the collection and analysis of data. This is already the goal of the H2020 RI-Paths project, as presented by A. Reid. This CSA action is currently ongoing and will help coming to a more homogeneous framework for impact assessment, needed for the next phase of the development of a particle collider-based research infrastructure

It is generally agreed that the realization of a new research facility is both a technological and a social process; the interests and expectations of the different participating actors show up in its design. This means that the economic and social impacts should no longer be generated by accident, in an episodic fashion but rather be inclusively identified and shaped, and then well integrated from the very start of the design process. This requires a new best-practice box of management tools, risk

[3]Abreu, M., Grinevich, V., Hughes, A., and Kitson, M. (2009), Knowledge exchange between academics and the business, public and third sectors, UK Innovation Research Centre, University of Cambridge and Imperial College London (http://www.cbr.cam.ac.uk/pdf/AcademicSurveyReport.pdf)

[4]Czarnitzki, D. and Thorwarth, S. (2012), "Productivity effects of basic research in low-tech and high-tech industries", Research Policy, 41(9), 1555–64

[5]González, X. and Pazó, C. (2008), "Do public subsidies stimulate private R&D spending?", Research Policy, 37(3), 371–389.

[6]Impact of CERN procurement actions on industry: 28 illustrative success stories, Centre for Industrial Studies, Milan (2018): https://cds.cern.ch/record/2670056

[7]Park, W.G. (1995), "International R&D Spillovers and OECD Economic Growth", Economic Inquiry, 33(4), 571–91.

analysis, administrative resources and strategy tuned to the size of the projects. Much as we devise new scientific instruments in relation to the precision we are aiming to achieve and the particle we want to study. Time will tell how these considerations can be applied to future research infrastructures and help increase their societal return. I hope that this workshop will motivate further interdisciplinary work to address these pressing questions.

Socio-Economic Impact Assessments of ESA Programmes: A Brief Overview

Charlotte Mathieu

Contents

1   The European Space Agency

The European Space Agency (ESA) is an intergovernmental organisation established in the seventies with today 22 Member States. Its purpose is to promote cooperation between Member States, in space research and technology as well as in space applications.

Among its 22 Member States,[1] 20 of them are EU Member States, the two others being Norway and Switzerland. In addition 7 other EU Member States have Cooperation agreements with ESA, and Slovenia recently became an Associate Member. Lastly, Canada has a special Cooperation agreement with ESA and participates in its programmes.

ESA's staff is working mainly on its 8 different sites in Europe. The Agency's budget is about €5.6 billion in 2019,[2] with €1.6 billion coming from the budget of the European Union, for the Galileo (Navigation) and Copernicus (Earth Observation) programmes and of EUMETSAT for meteorological satellites. The rest of the budget comes directly from the contributions of the Member States.

---

[1] See Fig. 1.

[2] See Fig. 2.

C. Mathieu (✉)
Industrial Policy and Economic Analysis Section, European Space Agency, Paris, France
e-mail: Charlotte.Mathieu@esa.int

© The Author(s) 2021   53
H. P. Beck, P. Charitos (eds.), *The Economics of Big Science*, Science Policy
Reports, https://doi.org/10.1007/978-3-030-52391-6_8

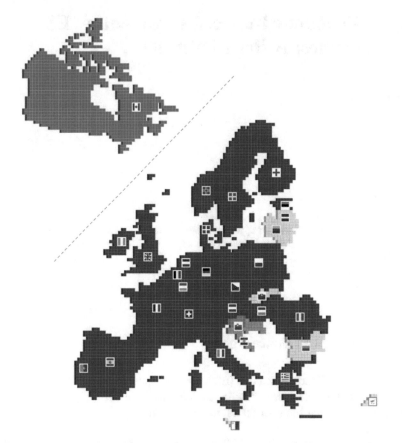

Fig. 1 Mapping of the 22 ESA Member States (in dark blue)s, Associate members (Slovenia), and States having Cooperation agreements (in grey)

About a fourth of the budget goes to the Earth Observation programmes, thus representing the largest set of activities in the Agency, followed by Launchers (Space Transportation), Navigation, and finally Human Spaceflight & Exploration).

ESA is one of the few space agencies in the world to combine responsibility in nearly all areas of space activity. Space science is a mandatory programme, i.e., all Member States contribute to it according to GDP. All other programmes are optional, funded "a la carte" by Participating States.

The Agency has designed, tested and operated over 80 satellites over the past 50 years. About 85% of ESA's budget is spent on contracts with European industry. ESA should ensure that Member States get a fair return on their investment.

ESA supports the development of the European space industry, which today, for the manufacturing part, sustains around 40,000 jobs. Europe is successful in the commercial arena, with a market share of telecom and launch services higher than the fraction of Europe's public spending worldwide. European scientific communities are world-class and attract international cooperation. European space research

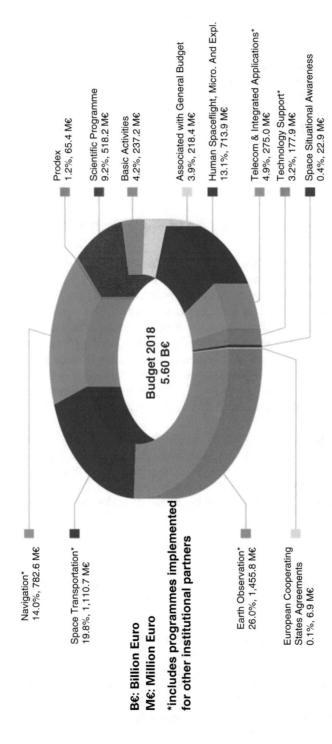

Prodex
1.2%, 65.4 M€

Scientific Programme
9.2%, 518.2 M€

Basic Activities
4.2%, 237.2 M€

Associated with General Budget
3.9%, 218.4 M€

Human Spaceflight, Micro. And Expl.
13.1%, 713.9 M€

Telecom & Integrated Applications*
4.9%, 275.0 M€

Technology Support*
3.2%, 177.9 M€

Space Situational Awareness
0.4%, 22.9 M€

Navigation*
14.0%, 782.6 M€

Space Transportation*
19.8%, 1,110.7 M€

Earth Observation*
26.0%, 1,455.8 M€

European Cooperating
States Agreements
0.1%, 6.9 M€

Budget 2018
5.60 B€

B€: Billion Euro
M€: Million Euro

***includes programmes implemented
for other institutional partners**

Fig. 2 ESA Budget for 2018, by domain

and innovation centres are recognised worldwide. European space operators (Arianespace, Eumetsat, Eutelsat, SES Global, etc.) are among the most successful ones in the world.

The ESA's governing body is the Council, in which each Member State is represented and has one vote. Every 3 years, the Council meets at the ministerial level ("Ministerial Council") to take the key decisions on the continuation of the programmes and the new ones coming as well as on their respective funding.

In 2019, the Council will meet in late November. ESA is therefore preparing proposals to the 22 Ministers in charge of space, complemented with an assessment of the socio-economic impact of its programmes.

2   The Socio-Economic Impact of Space Activities

Space is more and more part of our daily lives and has become an engine of economic growth and innovation with impacts that go well beyond its industrial sector [1]. ESA's Member States wish to ensure maximum benefits to the economy and society from their investments in space activities. The Agency, as any other modern public administration, has a responsibility to ensure the creation of value for society, in an end-to-end perspective (i.e. from technology research to service development). Governments need evidence that investing in space creates jobs and supports a competitive European economy of the future, while providing strategic tools to implement sovereign policies. ESA has therefore conducted more and more socio-economic impact assessment studies of its programmes over the past years to inform decision-makers.

In preparation of its last Council at Ministerial level in 2016, ESA conducted three studies on large programmes, including Ariane 5 and the International Space Station. This past couple of years, in preparation of the upcoming Council at Ministerial level, such studies were conducted not only ex post but also ex ante, on programmes of different sizes and in different domains, covering most of ESA's fields of activities.

3   Methodology of Socio-Economic Impact Assessment of ESA Programmes

There is no standard approach to assess the socio-economic impacts of space programmes in general and even more so to assess the impacts of all ESA programmes. Since 2012 ESA has consolidated its own methodological approach in line with recognised standards. The methodology of all the assessments conducted by ESA on its programmes was harmonised to the extent possible but it was also tailored to the mandate and strategic objectives of the programmes and respective stakeholders' interests. The specificities of the programmes imply differences in the scope and parameters of the impact assessments, be it the timeframe for the analysis, the availability of underlying data and data sources or an emphasis on certain types of impacts. For that reason, the results of the various assessments are independent for each study and a direct comparison between the results is only possible to a very limited extent.

The framework for assessing the impact of ESA programmes includes the definition of the impacts and indicators tailored to the programme, the definition of the assessment methodology of each indicator tailored to the programme (qualitative and quantitative), and an analysis of the sources of data and of the relevant stakeholders (incl. Scientific community, Member States, ESA, industry, research organisations and society). The impacts were assessed according to five categories: economic, scientific, technological, strategic and societal impacts.

As an example the **scientific impact** of an ESA programme would include:

- The volume and quality of the refereed papers based on ESA-led programmes and missions;
- The interest from the scientific community, through the number of users accessing ESA scientific data and the volume of data downloaded;
- The knowledge transfer and cross-fertilisation (when knowledge from scientific papers is used in other domains) as well as industrial cross-fertilisation (for example, refereed papers which originate from a collaboration between public and private partners);
- The international cooperation of scientists on ESA-led programmes and missions, with the measure of international co-authoring for scientific papers.

The **strategic impact** of an ESA programme would include:

- The level of international cooperation;
- The level of industry competitiveness, with for instance the number of patents;
- The level of European non-dependence.

The **societal impact** of an ESA programme would include:

- The level of public outreach through social media and visibility in the media;
- The contribution to education and careers in STEM, with for instance the number of PhD theses based on ESA-led missions;
- The contribution to sustainability, for instance with the space weather programmes contributing to the overall sustainability of space activities.

The **economic impact** of an ESA programme would include:

- The gross value added or the impact on the GDP;
- The number of jobs sustained and created;
- The support to innovation and technology transfers, including spin-offs and technological spill-overs.

To illustrate this approach, the assessment of the economic impact of ESA's R&D programme in Earth Observation, i.e. Future EO, concluded that:

- Every euro invested by ESA Member States in the programme creates €3.8 in their economy over 2013–2030, from which:
  - €1.9 corresponds to an increase in their GDP
  - €1.9 corresponds to longer term innovation spill-overs;
- More than 60% of the investment is recovered in tax revenues in the medium term;
- For each new job in the space sector, 1.3 additional jobs are created in the wider economy.

4   Conclusions

This paper gives an overview of the different types of impacts that ESA's space programmes have in different domains, measured with relevant and specific sets of indicators. The impacts of the future space missions to be proposed to the European Ministers at Space19+, the Council at Ministerial level that will take place in November 2019, will be evaluated to support their decisions. The main results will be available on ESA's Space Economy website (space-economy.esa.int).

ESA's programmes have demonstrated benefits as significant and diverse as other major scientific projects and the Agency will continue cooperating with other institutions to exchange on indicators and impact evaluation.

Acknowledgment   The author would like to thank Mr. Laurent Laurich (Analyst in Socio-Economic Impact Assessment studies, ESA, Paris, France) for his invaluable comments and discussions during the preparation of this essay.

Reference

1. https://space-economy.esa.int/initiatives/#first

Full Presentation

https://indico.cern.ch/event/727555/contributions/3461265/attachments/1868232/3072979/Presentation_FCC_Mathieu.pdf

Designing a Socio-Economic Impact Framework for Research Infrastructures: Preliminary Lessons from the RI-PATHS Project

Alasdair Reid

Contents

1 Introduction

The European Commission (EC)[1] defines research infrastructures (RIs) as facilities, resources and services that are used by the research communities to conduct research and foster innovation in their fields [1]. They include:

- Major scientific equipment or sets of instruments;
- Knowledge-based resources such as collections, archives or scientific data;
- E-infrastructures, such as data and computing systems and communication networks;
- Any other infrastructure of a unique nature essential to achieving excellence in research and innovation.

Contribution to the ECONOMICS OF SCIENCE—Workshop, Brussels, 25 June 2019.

---

[1] Article 2 (6) of Regulation (EU) No 1291/2013 of 11 December 2013: 'Establishing Horizon 2020 – the Framework Programme for Research and Innovation (2014–2020)'.

A. Reid (✉)
European Future Innovation System (EFIS) Centre, Louvain-la-Neuve, Belgium
e-mail: reid@efiscentre.eu; https://www.efiscentre.eu

© The Author(s) 2021
H. P. Beck, P. Charitos (eds.), *The Economics of Big Science*, Science Policy Reports, https://doi.org/10.1007/978-3-030-52391-6_9

This definition is now internationally accepted (e.g. by the OECD's GSF,[2] ESFRI[3] and the ERIC Council Regulation[4]). Moreover, RIs can be:

1. Single-sited facilities (unified single body of equipment at one single physical location);
2. Distributed facilities (a network of distributed resources: instrumentation, collections, archives, and scientific libraries);
3. Virtual facilities (e.g. ICT based system for scientific research, including high-capacity communication networks, and computing facilities providing services electronically, also known as e-infrastructures[5]), or
4. Mobile facilities (vehicles designed for scientific research).

2 The RI-PATHS Project

The growing importance placed on the role of RIs, both in Europe and globally, as a basis for conducting research, but also for research-based education, is evident from the increasing public funding allocated to the design, development, operation and upgrading/renewal of RIs. Under the EC's Horizon 2020 research and technological development framework programme (RTD FP), a key activity line is focused on research infrastructures and aims to "endow Europe with world-class research infrastructures that are accessible to all researchers in Europe and fully exploiting their potential for scientific advancement and innovation. The total budget for this activity in Horizon 2020 is €2.488 million.[6] In addition, during the period 2014–2020, the European Structural and Investment Funds (ESIF) are investing €6.931 million (out of a total planned expenditure by the European Union (EU) Member States of €10.131 million)[7] in public investments in favour of research and innovation infrastructures through the European Regional Development Fund (ERDF).

The overall scale of funding is thus significant and concerns investments in both European level RIs (ESFRI projects such as ELI, ESS, etc.) and national and regional scale RIs that serve a broad-based community of researchers and users (including business users, etc.). As the OECD [2] underline, the level and range of types of investments in RIs leads to a diverse group of stakeholders' (ranging from national and regional public authorities, funders including charitable foundations, RI managers, scientific and industrial users and special interest groups and ultimately

---

[2]Science Global Forum OECD (2014; 2016) and OECD (2014).

[3]ESFRI, the European Strategy Forum on Research Infrastructures (2010).

[4]Community Legal Framework for a European Research Infrastructure Consortium (ERIC) (2009).

[5]See the catalogue of e-infrastructure services developed by the eInfraCentral project available at https://www.einfracentral.eu/home

[6]See: https://ec.europa.eu/programmes/horizon2020/en/area/research-infrastructures

[7]Source data: https://cohesiondata.ec.europa.eu/d/s499-6d7x

civil society) with (differing) interests in the topic and outcome of RI impact assessments.

In this context, the RI-PATHS project[8] aims to provide policy makers, funders and RI managers the tools to assess RI impact on the economy and their contribution to resolving societal challenges, etc. The goal is to improve the understanding of the long-term impact pathways [3] of the various types of RIs operating in Europe, and, indeed, internationally. The project is being implemented over a 30-month period beginning in January 2018 and ending in June 2020. The activities include a review of the state of the art, consultation on the needs and practices of the RI community concerning IA, a phase of co-design with RI managers and stakeholders of an IA framework and the piloting of the IA framework by, at least, the four partner RIs (CERN, DESY, ALBA and ELIXIR). The project builds on and has ensured interaction with all relevant past and on-going initiatives including those by the OECD GSF, various ESFRI working groups, the ACCELERATE[9] project, etc.

Most approaches to date have sought to undertake specific type of impact assessments often at specific points in the lifecycle of an RI, including, e.g., cost-benefit analysis at the pre-investment stage or before an expansion or upgrading, impact on scientific excellence or the attraction of funding (from competitive programmes or industrial partners) during the operational phase, impact on suppliers through procurement, impact on human resources through training, doctoral research, etc. The RI-PATHS team acknowledge that a single 'one-fits-all-model' to assessing socio-economic impact is unlikely to meet the needs of the RI stakeholders. Rather the effort is focused on developing an IA framework adopting a modular approach with a generic core model and more detailed sub-models of IA. Moreover, the IA framework is designed to cover the whole life-cycle of RI development, operations and ultimately decommissioning.

The OECD [2] argued that impact assessment should be connected to the strategic objectives and mission of each RI. They proposed six standard objectives that they considered cover all the main dimensions of impact:

- Be a national or world leading scientific RI and an enabling facility to support science.
- Be an enabling facility to support innovation.
- Become integrated in a regional cluster/in regional strategies/Be a hub to facilitate regional collaborations.
- Promote education outreach and knowledge transfer.
- Provide scientific support to public policies.
- Provide high quality scientific data and associated services.
- Assume social responsibility towards society.

---

[8]See: www.ri-paths.eu. All deliverables of the project can be downloaded here.

[9]See: http://www.accelerate2020.eu/

3   A taxonomy for RIs

Taking account of the need to frame impact assessment in the context of the overall mission an RI has set itself (or in many cases has had set for it in the context of national, European or international funding), the RI-PATHS project developed [4] a taxonomy of types of RIs based on their mission and type of research conducted at the RI or research services provided to users (see Table 1).

The taxonomy was tested during the consultation phase of the project with RI stakeholders and managers. Overall, it was considered to be a useful framework as it covers a wide range of organisations with different objectives (60% of respondents of a survey of RI managers said it fits their case well). At the same, it was recognised that it may be difficult for many RIs to place themselves into just one category. Hence, the typology served as a framework (during the participatory workshops run by the project) to test the hypothesis that there are distinct pathways to impact depending on the level of interaction with users. To this end, a broader distinction can be made between:

(a) RIs that offer services and perform own research activities;
(b) RIs that are exclusively or primarily user/services oriented.

In line with OECD [5], we define impact as any long-term effect, whether intended, unintended, positive, negative, direct or indirect produced by an intervention. They are the ultimate changes produced in the society by means of a given action or investment decision. Having an overarching list of impact categories can allow a RI to identify where it is not having impacts, as well as where it is having some. This is an important issue for accountability and advocacy, since it allows RIs to investigate more closely why, for instance, it is 'failing' to achieve desired results under certain types of impacts.

A distinction can be made between impacts caused by the RI pursuing its core science mission and impacts caused by the RI being a socio-economic actor. In broad terms, these are both socio-economic impacts resulting from the operation of an RI. The former can be e.g. contribution to short-term and/or long-term problem solving, qualification of scientists, impacts on innovation and productivity in the economy, opening up of new perspectives in the policy discourse and outreach and popularisation of knowledge in society. Examples of the latter include employment effects, wages paid and multipliers, qualified procurement effects with impact on innovation, procurement of standardised, off-the-shelf goods and qualification effects for technical staff.

Demonstrating the causal pathway between investment in RIs and the wider socio-economic impacts can be an intricate, complex and labour-intensive task. This suggests the need for a framework that classifies the diverse impact pathways. An IA framework should provide the opportunity to visualise the potential pathways to impacts or the forms of impact that exist. Visualisation can also help understand the system by which impacts are produced and thus improve it.

Table 1 Taxonomy of RIs by mission/type of research conducted

Type of research/ service	Definition	Examples of RIs
Pure fundamental/basic research	Curiosity-driven research that advances human knowledge. Generating socio-economic impact potentials are not the priority	Mission-led facilities advancing knowledge often on the fundamental understanding of nature/universe, without immediate or even medium-long term prospect of practical application. They are usually complex and capital-intensive facilities. Examples include CERN in Switzerland, the square kilometre Array (SKA); the Atacama large millimetre Array (ALMA); Facility for Antiproton and ion Research (FAIR) in Germany
Use-inspired basic research	Scientific research conducted with the clear ambition of solving known societal challenges or creating technologies for future economic applications	Use-inspired research lies between fundamental research and applied research. This category includes facilities and organisations whose main objective is to increase scientific knowledge for the direct benefit of humankind and the ecosystem. The problems addressed are of a practical nature. Examples include ICOS ERIC, the integrated carbon observation system (head office in Finland), MaRINET in the field of renewable energy
Applied and solution-oriented research	Research and development directly aimed at meeting public or business demands and at responding to well identified research or technological problems.	Solution-oriented research facilities pursue defined contributions with potential application. Their core mission is to deliver technologically advanced services to potential users (citizens or firms) involving practical application of science and motivated by the need to solve immediate problems. Such facilities also offer contract-based research. Includes competence centres and labs specialised in a particular field (e.g. clinical research centres, laser light facilities, etc.). Examples include the high field magnet Laboratory in Netherlands, the multidisciplinary seafloor and water-column observatory (EMSO), the European clinical research infrastructure network (ECRIN), the National Centre of oncological Hadrontherapy (CNAO) in Italy.

(continued)

Table 1 (continued)

Type of research/ service	Definition	Examples of RIs
Facilities providing scientific services	Facilities designed to offer services to be directly used by the science community to efficiently carry out their research	Includes innovation centres, centres for experimental development, design centres, facilities and equipment for developing and testing prototypes and innovation not yet intended for commercialisation, data repositories. Examples include the diamond light source in UK, the ELIXIR data infrastructure, the common language resources and technology infrastructure (CLARIN), the Partnership for Advanced Computing in Europe (PRACE)

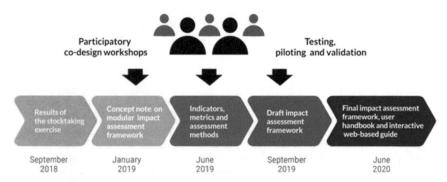

Fig. 1 Timeline of the RI-PATHS project

Accordingly, a core element of the RI-PATHs project has been on co-creating with the RI community a logical model that links RI activities to outputs, outcomes, longer-term results and finally impacts [6]. As described in Fig. 1, the IA framework has been developed in two stages via a series of participatory co-design workshops, followed by a validation workshop and piloting in the four partner RIs before the launch of a final impact assessment framework.

4 Discussing Preliminary Results

The findings of the first round of workshops underlined that there are three main lines of inquiry relevant for all types of RIs (even if each of them may be more or less strongly associated with a specific type).

- Quantifiable impacts that can be captured through quantitative metrics available e.g. in economic analysis: In some areas, impacts can be determined reliably based on economic modelling. Areas to be thus covered include the transmission of socio-economic effects through employment, (qualified) procurement, some dimensions of training as well as RIs 'gravitational effect' on visitors. This approach takes and justifies assumptions but does not further reflect on specific pathways in the process. It has the clear advantage to produce aggregate figures in monetary terms and easy-to-process results for the political discourse.
- Non-quantifiable impacts the assessment of which can, however, be supported by quantitative means (e.g. social network analysis, dedicated surveys): Some impacts can be characterised through quantities but not suitably translated to monetary terms and aggregated. In these areas actor-based analysis can chart patterns of interaction between RIs and their socioeconomic environment, most prominently their users. It can quantify these interactions and to the extent possible attribute specific effects to them. Thus, it can help establish not only the socio-economic, but also the societal contribution of RIs based on the properties of the network that they have built around them and the known impact pathways that they can be demonstrated to trigger.
- Complex network effects captured through exploratory approaches which require qualitative approaches such as narrative-based case studies: The analysis of network effects has been highlighted as an area of particular relevance for distributed RIs. It has so far not been intensively explored and will require more work to establish methodological foundations. Ideally, it should be possible to outline a core set of quantitative indicators for this area, e.g. in the area of human capital. In many other areas, this type of analysis will look at complex issues that quantitative approaches fail to satisfactorily capture.

During the first round of RI-PATHS participatory workshops four categories of impacts were used to frame the discussions: (1) Direct economic impact, (2) Human resources impact, (3) Social/societal impact and (4) Policy impact (see Fig. 2). During the workshops, these four broad categories were used to identify twelve stylised impact pathways that help to measure the multi-faceted impact of research infrastructures [7].

The goal of the second round of workshops was to engage the RI community in the identification of suitable indicators and metrics that can, in practice, serve to measure the various impacts identified in the first round. As a result of the discussions, the RI-PATHS team decided to re-phrase the initial set of pathways as follows:

Impacts as a result of RIs Enabling Science (i.e. performing their primary mission)

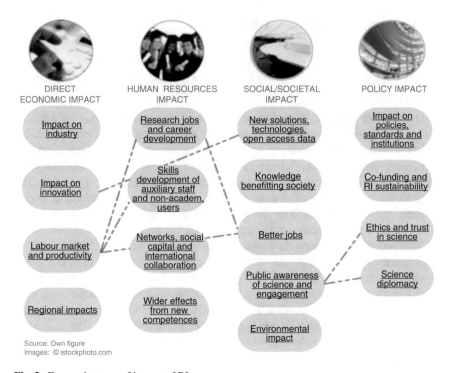

Fig. 2 Four main types of impact of RIs

- Employment & standardised procurement
- Publication-citation-recognition (incidental spillovers)
- Technology transfer & licensing (supported spillovers)
- Learning- and training-through-specialised procurement
- Generic learning- and training-through-usage

Impacts as a result of RIs Supporting Problem Solution (by providing services to users)

- Interactive industrial problem solution
- Interactive public sector problem solution
- Provision of specifically curated/edited data to industry
- Provision of specifically curated/edited data to public sector
- Other interactive societal problem solution

Impacts through shaping future trends in Science and Society

- Changing fundamentals of research practice
- Contribution to formal standards in science
- Creating and shaping scientific networks and communities
- Creating and shaping networks between science and society
- Communication, outreach and engagement.

5 Conclusion

The proposed conceptual framework (see Fig. 3) fulfils two central functions:

- First, it defines impact domains where RI may play a role. While political expectations may shift, findings from the participatory workshops and surveys have confirmed that IA should no longer be solely motivated from a purely compliance-oriented perspective. Increasingly, many RIs are articulating their mission to contribute to broader societal goals.
- Second, the model helps to define the logic of impact causation, or impact pathways, through which different types of impact materialise. Understanding these pathways can assist decision making by RI managers in terms of the launch of or intensification of specific internal activities or through the collaboration and engagement with external partners.

Given this overall framework, the second round of workshops were also used to assess and refine a long-list of indicators down to a workable set that can be used to track the various types of impacts and the inter-relations between the impact pathways [8]. In the final step of the project, this approach will be piloted, then refined into an operational guidance that will provide RI managers and stakeholders with an easy-to-use framework that helps them identify the sort of questions an IA exercise should answer and what data needs to be collected.

Moreover, the EC has stated that the RI-PATHS framework will guide the future IA of ESFRI projects and the project team expect that the guidance (to be licenced under Creative Commons) will be further developed by practitioners in the coming years.

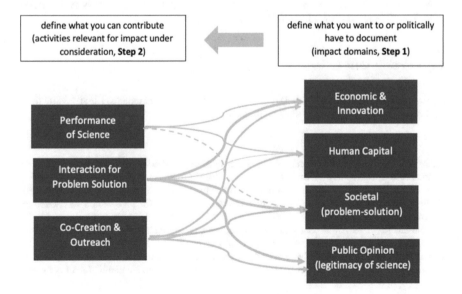

Fig. 3 General conceptual approach of the RI-PATHS Framework

Acknowledgements This short paper summarises the work carried out by the RI-PATHS consortium partners to mid-2019 and draws on the key deliverables coordinated, notably, by Eline Griniece (EFIS Centre), Henning Kroll (Fraunhofer ISI) and Silvia Vignetti (CSIL).

References

1. EU (2013) Consolidated text: Regulation (EU) No 1291/2013 of the European Parliament and of the Council of 11 December 2013 establishing Horizon 2020 - the Framework Programme for Research and Innovation (2014-2020) and repealing Decision No 1982/2006/EC (Text with EEA relevance). https://eur-lex.europa.eu/legalcontent/EN/ALL/?uri=CELEX:02013R1291-20150704
2. OECD (2019), "Reference framework for assessing the scientific and socio-economic impact of research infrastructures", OECD Science, Technology and Industry Policy Papers, No. 65, OECD Publishing, Paris, https://doi.org/10.1787/3ffee43b-en.
3. Griniece, E., Reid, A. and Angelis, J. (2015). Guide for Evaluating and Monitoring the Socio-Economic Impact of Investment in Research Infrastructure. DOI: https://doi.org/10.13140/RG.2.1.2406.3525/1
4. Giffoni, F, Vignetti, S., Kroll, H., Zenker, A., Schubert, T., DeYoung Becker, E., Ipolyi, I., Griniece, E., Angelis, J., (2018). Working note on RI typology. RI-PATHS Project (Brussels) DOI: https://doi.org/10.13140/RG.2.2.29020.23684
5. OECD (2002) Glossary of Key Terms in Evaluations and Results Based Management. Re-printed in 2010
6. Vignetti, S., Griniece, E., Cvijanovic, V., Reid, A., Helman, A., Kroll, H. (2019), Report on stocktaking results and initial IA model. RI-PATHS project (Brussels). DOI: https://doi.org/10.13140/RG.2.2.34844.95360
7. Griniece, E., Kroll, H., Cvijanovic, V., Zenker, A., Reid, A. (2019) Concept note of the modular impact assessment framework. RI-PATHS project (Brussels) DOI: https://doi.org/10.13140/RG.2.2.28102.42566
8. Kroll, H., Zenker, A., Hansmeier, H., Griniece, E., Helman, A., Angelis, J, Vignetti, S., Reid, A. (2019) Consolidated report on the participatory workshop results. RI-PATHS Project (Brussels). DOI: https://doi.org/10.13140/RG.2.2.34813.31201

Full Presentation

https://indico.cern.ch/event/727555/contributions/3461264/attachments/1868059/3072628/190625_RI-PATHS_FCC_workshop_Reid.pdf

Findings from the LHC/HL-LHC Programme

Andrea Bastianin

Contents

1 Introduction

CERN and more generally Big Science Centres (BSCs) are ideal testing grounds for theoretical and empirical economic models. In fact, the operations of BSCs generate unique data for economists (about e.g. procurement contracts, staff, students and alike, software, media coverage). See Castelnovo et al. [1] and references therein. Moreover, governance and procurement policies of BSCs are interesting topics in management studies (see e.g. [2, 3]). Furthermore, innovation and breakthrough technologies arising from BSCs are one of the drivers of long-run economic growth [4–6]. Finally yet importantly, as shown in Fig. 1, CERN and its accelerator complex represent a unique international research infrastructure (RI) that generate a variety of societal benefits that go well beyond the boundaries of the scientific community using them for research purposes. See Florio [7] and Florio and Sirtori [8] for recent overviews of the CBA of RI. Boardman et al. [9] is a useful general introduction to the methodology.

A social CBA is an appropriate methodology for evaluating a RI because it translates in quantitative terms the multi-dimensional benefits ascribed to it [10]. The social CBA machinery has found applications in several policy-relevant

A. Bastianin (✉)
Department of Economics, Management, and Quantitative Methods (DEMM), University of Milan, Milan, Italy
e-mail: andrea.bastianin@unimi.it

H. P. Beck, P. Charitos (eds.), *The Economics of Big Science*, Science Policy Reports, https://doi.org/10.1007/978-3-030-52391-6_10

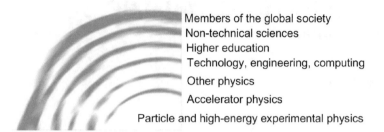

Members of the global society
Non-technical sciences
Higher education
Technology, engineering, computing
Other physics
Accelerator physics
Particle and high-energy experimental physics

Fig. 1 The "onion" model of involvement: societal benefits of particle accelerator. Source: http://cds.cern.ch/record/2653673

contexts. For instance, successfully passing a CBA test is required for co-financing major projects with the European Regional Development Fund and the Cohesion Fund. Similarly, the "Horizon 2020—Work Programme 2018–2020 on European research infrastructures" mentions that the preparatory phase of new ESFRI projects (www.esfri.eu) should include a CBA [11]. Assessment of the socio-economic impacts of RIs has been included in the latest edition of the "Guide to Cost-Benefit Analysis of Investment Projects" of the [12].

2 Methodology

One of the main object of interest in a social CBA is the estimated expected Net Present Value (*NPV*) of a RI at the end of a defined observation period:

$$E(NPV) = E[DB - DC] \tag{1}$$

Interpretation of results is straightforward: a RI passes the CBA test when the cumulative sum of discounted social benefits (*DB*) exceeds the cumulative sum of discounted social costs (*DC*), that is when the expected *NPV* is greater than zero. See Bastianin and Florio [13] for further details.

The practical implementation of model (1) involves the following steps:

1. Identification of the social benefits and costs that are relevant for the HL-LHC;
2. Estimate present and future social benefits and costs;
3. Use Monte Carlo simulations to estimate the probability distribution of social costs, benefits and of the NPV of the project.

Since the HL-LHC represents an upgrade of an existing RI—the LHC—we carry out the CBA for two scenarios. The baseline scenario is CERN with the HL upgrade and the counterfactual scenario (CFS) is the operation of the LHC until its end of life without the HL upgrade. The horizon of the analysis spans the 1993–2038 period. From 1993 to 2014, the two scenarios overlap and hence the costs and benefits are

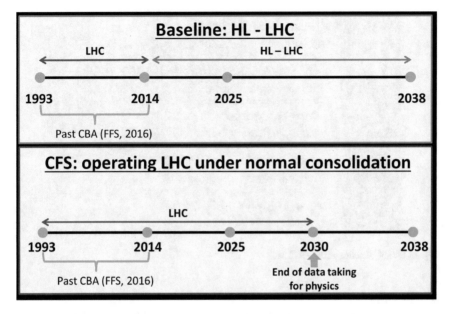

Fig. 2 HL-LHC and Counterfactual scenario: 1993–2038. Notes: dates in the figure represent time schedules that are consistent with assumptions from discussions with CERN experts

identical to those considered in the CBA of the LHC by Florio, Forte and Sirtori [14] (see Fig. 2).

In the CFS the LHC is operated "with ordinary consolidation activities"; after 2031, data taking ends and CERN staff shift their engagement to other scientific activities. After the collider is switched-off, the equipment remains in the tunnel and the underground infrastructure would be subject to appropriate monitoring and safety procedures without operating. Planned maintenance and repair activities are considered. In both scenarios we consider the following as the most relevant benefits:

1. The value of training (or human capital formation) for students and early stage researchers
2. Technological or industrial spillovers for collaborating firms and other economic agents
3. Cultural effects for the public
4. Academic publications and pre-prints for scientists
5. Existence or public good value of the RI for non-users

Notice that although the horizon of the analysis spans the years 1993–2038 some benefits extend over this time period (see Fig. 3). Non-users are people who currently do not directly use the services of the RI, but are better off simply knowing that new knowledge might be created.

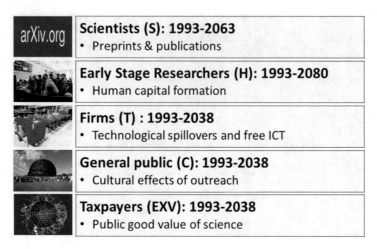

Fig. 3 Social benefits of the HL-LHC

3 Results

In what follows we consider the incremental benefits produced by the HL-LHC, namely the cumulated discounted value of the benefits associated with the HL-LHC minus the corresponding cumulated discounted value arising in the CFS.

For brevity, we focus on three categories of benefits. There are two reasons for restricting our attention to these categories: First, they are quantitatively the most relevant. Second, CERN can—to some extent—"control" these benefits and therefore they represent a strategic lever for promoting future RI.

Benefits for students and early stage researchers measure the salary increase or premium over the entire work period for individuals that have been involved in the LHC program. This is the single most important benefit; in fact, it represents 40% of total incremental benefit due to the HL-LHC. Industrial spillovers arise for firms working with CERN by resulting in new products, services, creation of new business opportunities and more efficient operation for companies. Overall, benefits in this category account for 37% of the total incremental benefit generated by the HL-LHC scenario. Although we have considered a variety of cultural effects, most of the benefits in this category are due to onsite CERN visitors and visitors of CERN travelling exhibitions. Cultural benefits account for 5% of the total incremental benefit.

Comparing the benefits and costs, shows that the Net Present Value of HL-LHC is positive and greater than the one of the counterfactual scenario. The difference between the NPVs of the two scenarios is the direct benefit of the HL-LHC project, since the alternative scenario is to continue the operation of the LHC only until the end of its lifetime with no further operation. The ratio between the HL-LHC and CFS total cost difference and the HL-LHC and CFS total benefit difference is 1.7. This implies that every CHF spent on the HL-LHC project generates 1.7 CHF of benefits for the society.

4   Monte Carlo Analysis

One key step in a CBA is the Monte Carlo analysis that relies on simulations to deliver an estimate of the uncertainty of costs and benefits of the RI. Uncertainty arises not only because of the long time horizon of the CBA, but also because in the construction of scenarios for costs and benefits we rely on a set on a set of unknown parameters that can be estimated from the data, selected from previous studies or guess-estimated with the consultation of experts. These parameters, data and proper formulas lead to an estimate of the yearly value of costs and benefits. Drawing at random these parameters from appropriate statistical distribution functions and repeating the CBA a large number of times allows to estimate the empirical distributions of social costs, benefits and of the NPV.

Application of this methodology to the HL-LHC scenario yields an estimate of the probability of obtaining a negative NPV equal to 13%. It is fair to report that the CBA of the HL-LHC is based on a set of very conservative assumptions about some of the benefits that have probably led to under-estimating them.

5   Conclusions

In view of the application of the CBA methodology to other RI at CERN or in other BSCs it is important to stress both its merits and its limits. Results of a social CBA do not depend on the scientific utility of a RI, nor can they be used to rank different RI based on their discovery potential.

What a successful CBA does is to identify and quantify the most relevant social costs and benefits related with a RI. The attractiveness of CERN for Early Stage Researchers (ESR) is key for passing the social CBA test. Relations with firms in the supply chain, development of Information and Communication Technologies (ICT), and cultural effects—especially those related to onsite visitors—are additional strategic levers that CERN could use to boost the societal benefits. A further aspect that is crucial for the improvement of CBA is data collection during the daily operation of BSC. Better data leads to more accurate and reliable estimates of social cost and benefits. Therefore, we identify data collection as a key factor to improve the CBA of RI.

Acknowledgments The analysis underlying this note has been prepared as a contribution to the FCC study (DCC-GOV-CC-0004, EDMS 1390795) in the frame of the Collaboration Agreement between the University of Milan and CERN (KE3044/ATS). Several people at CERN have contributed to the analysis underlying this short note. Special thanks go to Panagiotis Charitos, Johannes Gutleber Lucio Rossi and Florian Sonneman for insightful discussion.

References

1. Castelnovo P., Florio M., Forte S., Rossi L., & Sirtori, E. (2018). The economic impact of technological procurement for large-scale research infrastructures: Evidence from the Large Hadron Collider at CERN. Research Policy, 47(9): 1853–1867.
2. Florio M., Giffoni F. (2018). Scientific Research at CERN as a Public Good: A Survey to French Citizens. CERN-ACC-2018-0024. Available online at: https://cds.cern.ch/record/2635861.
3. Vuola O., Hameri A.P. (2006) Mutually benefiting joint innovation process between industry and big-science. Technovation 26(1): 3–12.
4. Breshnahant T., & Trajtenberg M. (1995). General purpose technologies: engines of growth?" Journal of Econometrics, 65(1): 83–108
5. Crépon B., Duguet E., and Mairesse J. (1998). Research, innovation and productivity: an econometric analysis at the firm level. Economics of Innovation and New Technology, 7(2), 115–158.
6. Helmers C. & Overman H. G. (2017). My precious! The location and diffusion of scientific research: evidence from the Synchrotron Diamond Light Source. The Economic Journal, 127 (604), 2006–2040.
7. Florio M., (2019) Investing in Science. Social Cost-Benefit Analysis of Research Infrastructure, MIT Press.
8. Florio, M., & Sirtori, E., (2016) "Social benefits and costs of large scale research infrastructures." Technological Forecasting and Social Change, 112, 65–78.
9. Boardman, A. E., Greenberg, D. H., Vining, A. R., & Weimer, D. L. (2017). Cost-benefit analysis: concepts and practice. 4th Edition. Cambridge University Press.
10. Florio M., Bastianin A., Castelnovo P. (2018a). "The socio-economic impact of a breakthrough in the particle accelerators' technology: a research agenda", Nuclear Instruments and Methods in Physics Research Section A, 909, 21–26.
11. European Commission (2017). Horizon 2020 – Work Programme 2018–2020. European research infrastructures (including e–Infrastructures). Annex 4, European Commission Decision C(2017)7124 of 27 October 2017. Available online at: https://ec.europa.eu/program.
12. European Commission (2014). Guide to Cost-Benefit Analysis of Investment projects. DG Regional and Urban Policy.
13. Bastianin, A. & Florio, M. (2018). Social Cost Benefit Analysis of HL-LHC, CERN Technical Report, CERN-ACC-2018-0014. Available online at: https://cds.cern.ch/record/2319300.
14. Florio, M., Forte, S., & Sirtori, E., (2016) Forecasting the socio-economic impact of the Large Hadron Collider: A cost–benefit analysis to 2025 and beyond." Technological Forecasting and Social Change, 112, 38–53.

Full Presentation

https://indico.cern.ch/event/727555/contributions/3461267/attachments/1868055/3072623/Bastianin_Findings_from_the_LHC.pdf

Designing a Research Infrastructure with Impact in Mind

Silvia Vignetti

Contents

1 Motivation

Assessing the socio-economic impact of a Research Infrastructure (RI) requires an evaluation framework and a data collection plan that should be put in place as early as possible. This essay maintains the idea that if data collection for impact assessment is not episodic, motivated by external requests from stakeholders and funding agencies, but rather becomes a routine activity, it could maximize its potential to inform the RIs management and serve the strategic planning function of the infrastructure.

2 An Increasing Demand of Socio-Economic Impact Assessment

There is an increasing demand of demonstrating the expected socio-economic impacts triggered by research infrastructures already during the preparatory phase. Traditionally, the case for funding a new or upgraded RI has been usually advocated by scientific collaborations often supported by peer reviews exercise on the basis of a

S. Vignetti (✉)
Development & Evaluation Studies, Centre for Industrial Studies, Milan, Italy
e-mail: vignetti@csilmilano.com

© The Author(s) 2021
H. P. Beck, P. Charitos (eds.), *The Economics of Big Science*, Science Policy
Reports, https://doi.org/10.1007/978-3-030-52391-6_11

scientific case [1]. More recently, ex-ante impact assessment is becoming a routine analysis complementing the scientific and business cases accompanying the request for public funding (see for example the ESFRI Roadmap Guide 2021 and the Proposal Submission Questionnaire) [2, 3].

The fact that RIs can demonstrate their contribution to socio-economic development beyond the primary objective of excellent science is becoming crucial in the increasing international competition of hosting countries [4, 5]. The importance of promoting science facilities as a hub of knowledge and innovation within a wider ecosystem with a relevant territorial dimension is fully embraced by the EU Cohesion Policy. Investments funded by the European Structural and Investment funds for RIs should be planned however in the context of wider national or regional innovation strategies (according to the smart specialisation approach).

Against this background, evidence from a recent survey collecting almost 200 questionnaires from RI managers in the context of the Ri-Paths project,[1] indicates that 40% of respondents have some experience with socio-economic impact assessment [6]. However, only half of them carry out impact assessment on a regular basis, while the remaining ones only episodically and upon request. The main motivation is the need to comply with formal procedures for funding applications. Internal initiatives include mostly the need for strategic evaluation and reporting rather than ensuring public accountability. In addition, in the majority of cases, the socio-economic impact assessment has been carried out internally with own resources and only in few cases externally contracted to professional experts.

More widespread is the practice of performance monitoring as a routine reporting tool to funding agencies for accountability purposes. When looking at existing practices for data collection and monitoring based on respondents' experience, the survey found that 41% collect data for management and compliance purposes based on a list of key performance indicators (KPIs), while only 9% collect impact indicators for the purpose of impact assessment exercises.

Impact assessment is linked but not identical to monitoring: performance monitoring is rather a continuous process generating data to track the progress of an action while impact assessment is a structured process that takes place at a given point in time allowing to assess the implications (past, future or both), of proposed actions. Monitoring indicators can be useful input to impact assessment exercises but shall not be confused with impact indicators.

Many RI managers reported that a systematic procedure to ensure efficient data collection and monitoring is not done in the majority of cases and among the main barriers to perform such activities there is the lack of internal resources and expertise. As a matter of fact, especially for large and multi-purpose facilities, there is the need for deep understanding and collaboration of different managers in the same RI to engage in a systematic data collection activity.

---

[1]https://ri-paths.eu

3   A Plan for Impact Assessment

Setting up a plan for impact assessment means deciding the roles (who does what), the timing (when: ex-ante, in itinere, ex-post, routinely every a number of years, etc.), the types of impact that are assessed (one or a sub-set in particular, all of them), which methods and techniques are used (with the related data collection methods and tools) and what is the expected use of the results in terms of feedback on strategy and management.

Currently there is no unique methodological framework for socio-economic assessment of RI but a variety of methods are used depending on the scope of the analysis, the type of impacts and the target users. Methods range from macroeconomic modelling to cost-benefit analysis to more qualitative narratives and case studies [7].

Despite different methodological approaches there is some consensus about the main impact areas (see [8–12]):

- Scientific impact
- Education impact
- Technological spillover and innovation
- Cultural and outreach
- Science as a public good

For each of these items a short list of common indicators and tools for data collection can be identified (see Table 1). The list can be adjusted and expanded/restricted according to the specific need and remit of the assessment as well as the nature and type of activities of each RI.

As can be seen from the table below systematic tracking of scientific publications and citations, procurement contracts, patents and other innovation output, visitors and doctoral students, social media output and other dissemination products are the basis for a solid assessment and the condition for evaluability. In addition, regular surveys to former students, supplier companies and other users help to grasp useful insights on the way impacts materialise and how to maximise them. They require data collection strategies and evaluation plans. If a dedicated data collection plan is not in place, the ex-post reconstruction of past data can be challenging if not totally unfeasible. For example, getting in contact with students and researchers who spent time at a RI some years before might be impossible. Data protection issues are relevant in a number of cases, calling for a proper assessment and a rigorous management plan to ensure that there is no infringement to existing legislation.

For these reasons, setting up and implementing a plan for impact assessment from an early phase can make data collection more efficient and effective, despite some initial design costs.

Table 1 Examples of common impact indicators and tools for data collection by main impact areas

Impact area	Indicators	Tools for data collection
Scientific impact: The value of knowledge and its dissemination	• N. of scientific publications (in impacted/peer-reviewed journals, periodicals, technical reports, . . .). – Of authors/scientists from the RI or – Scientists using the RI • N. of citations (track the wave of knowledge dissemination) • N. of attendees to conferences, workshops, seminars – Origin and duration of stay – Travel costs • Time needed to produce/use scientific outputs • Yearly salary of scientists	• Mandatory citation system • Tracking system on existing databases (web of science, Scopus, PubMed, arXiv, INSPIRE, etc..) based on word search
Scientific impact: Data and ICT	• N. of (FAIR) data content, open source data/software • N. of users/downloads • Time spent in producing data or ICT/time saved to reproduce or process data • Yearly salary of scientists	• Mandatory citation system for data and ICT tools • Users surveys (users' communities) • Tracking of downloads
Education impact	• N. of early career students/technical staff – Origin and destination – Skills acquired – Short-term/long-term • N. or attendees to trainings, workshops, summer schools by origin and duration of stay • Travel costs • Salary over a lifetime career • Salary of a control group	• Tracking system of students/alumni • Systematic surveys to track career paths and wage development • Systematic surveys to control group (ethics and data protection management)
Technological spill over and innovation	• N. of industrial suppliers – Value of contract – Year of contract – Technological classification of contract (high-medium-low) • N. of industrial users or collaborative projects with industry • N. of spin-offs/start-ups • Survival rate of spin-offs/start-ups • Incremental profit for new products/services/processes • Patents • Patents citations (backward and foreword)	• Systematic surveys to companies (with control groups) • Systematic surveys to start-ups and spin-offs • Analysis of balance sheets (e.g. Orbis database) • Tracking of patents (e.g. PATSTAT)

(continued)

Table 1 (continued)

Impact area	Indicators	Tools for data collection
Cultural and outreach	• N. of physical and virtual visitors • N. of events, communication and dissemination products and related users – Origin and duration of stay for physical visitors – Travel costs • Time spent for virtual visits (website/social media)	• Tracking n. of visitors • Survey to visitors • Media tracking • Web analytics
Science as a public good	• Contribution by member states • Taxpayers by member states • Willingness to pay for science	• Surveys to citizens to assess their willingness to pay

4 Conclusion

Socio-economic impact assessment of research infrastructure is a challenging and time-consuming activity. Yet, RI managers are increasingly faced by requests by funding agencies and policy makers to show the socio-economic impacts beyond the purely scientific one. For this reason, the collection of a well identified list of impact indicators shall be planned and implemented early in advance to avoid problems of unavailable or too expensive data to be produced on past actions. Including impact data collection within the routine activities for strategic management, not only will make the assessment exercise more efficient and effective, but would also increase its solidity and thus credibility as a strategic management tool.

References

1. Pancotti, C., Pellegrin, J. and Vignetti, S., 2014, "Appraisal of Research Infrastructures: Approaches, methods and practical implications", Department of Economics, Management and Quantitative Methods University of Milan Working Paper n. 2014-13.
2. ESFRI Roadmap Guide 2021, 2019.
3. ESFRI, Proposal Submission Questionnaire, 2018.
4. Macilwain, C., 2010. Science economics: what science is really worth. Nature 465, 682–684.
5. Scaringella, L., Chanaron, J., 2016. Grenoble-GIANT territorial innovation models: are investments in research infrastructures worthwhile? Technological Forecasting and Social Change, 112: 92–101
6. Catalano, G., Vignetti, S., Ipolyi, I., DeYoung Becker, E. and Dostalova, Z., (2018). Survey of Research Infrastructures on impact assessment practices, RI-PATHS project, Brussels, https://doi.org/10.5281/zenodo.3946309
7. Giffoni, F., Vignetti, S., 2019, Assessing the Socioeconomic Impact of Research Infrastructures: A Systematic Review of Existing Approaches and the Role of Cost-Benefit Analysis, L'industria, Fascicolo 1, gennaio-marzo.

8. Giffoni, F., Schubert, T., Kroll, H., Zenker, A., Griniece, E., Gulyas, O., Angelis, J., Reid, A. and Vignetti, S., (2019), Literature review on socio-economic impact assessment of Research Infrastructures, RI-PATHS project, Brussels, https://doi.org/10.5281/zenodo.3946298
9. Florio, Massimo, *Investing in Science. Social Cost-Benefit Analysis of Research Infrastructures*, MIT Press, 2019
10. Martin, Ben R. 1996. "The use of multiple indicators in the assessment of basic research." *Scientometrics*, 36(3), 343–362.
11. Martin, Ben R., and Puay Tang. 2007. *The Benefits from Publicly Funded Research*. Science Policy Research Unit, University of Sussex, Brighton, UK.
12. Salter, A.J., Martin, B.R., 2001. The economic benefits of publicly funded basic research: a critical review. Res. Policy 30 (3), 509–532.

Full Presentation

https://indico.cern.ch/event/727555/contributions/3461268/attachments/1867486/3071404/FCC_Week_DEF.pdf

Leveraging the Economic Potential
of FCC's Technologies and Processes

Linn Kretzschmar

Contents

An international consortium of more than 150 organizations worldwide is studying the feasibility of various future particle colliders to expand our understanding of the inner workings of the Universe. At the core of the Future Circular Collider (FCC) study is the design of a 100 km long circular particle collider infrastructure that could extend CERN's current accelerator complex with an integral research program that spans 70 years. The first step would be an intensity-frontier electron-positron collider allowing to study with precision the Higgs couplings with many of the Standard Model particles and search with high-precision for new physics while the ultimate goal is to build a proton collider with a c.m.s collision energy seven times larger than the Large Hadron Collider. Hosted in the same tunnel and profiting from the new infrastructure, FCC-hh would allow to explore a new energy regime where new physics may be at play.

One of the technologies lying at the heart of energy-frontier colliders is super-conductivity. The dipole magnets for steering the charged particles in their circular trajectories call for superconducting magnet technology that remains yet to be developed while this technology should be available at large scales as required by the size of a new collider-based RI. Since it takes decades for such a technology to reach industrial maturity, research and development efforts are well underway.

In this context and starting in 2017, the Institute for Entrepreneurship and Innovation at the Vienna University of Economics and Business conducted three research projects to analyze the potential of new application fields outside of the particle collider domain for superconducting magnets and important steps of their

L. Kretzschmar (✉)
MSCA EASITrain Fellow, University of Economics, Vienna, Austria
e-mail: linn.kretzschmar@wu.ac.at

© The Author(s) 2021
H. P. Beck, P. Charitos (eds.), *The Economics of Big Science*, Science Policy
Reports, https://doi.org/10.1007/978-3-030-52391-6_12

manufacturing value chain [1]. The overarching question behind these research projects is how, apart from learning more about the universe, society can benefit from a new research infrastructure during the design, construction and operation phase. In order to exploit the value of superconducting technologies utilized to build an electron-positron collider for other industrial purposes, the knowledge acquired from manufacturing these technologies needs to be spread and appropriated. To achieve this goal, the Institute of Entrepreneurship and Innovation at Vienna University of Economics (WUW), under the lead of CERN, identified innovative application fields outside of the particle collider domain for superconducting magnets the processes involved in manufacturing those technologies and assessed these new application fields and conducted elaborate market analyses for the most promising ones. For a detailed description of the projects, see [2] and [3]. The following paragraphs outline the methodology and main findings of these projects.

To identify new market opportunities for superconducting magnets and its manufacturing steps, the manufacturing value chain needed to be analyzed with regard to their importance. Based on factors such as uniqueness and cost intensity, the superconducting Rutherford cable, the furnace for the thermal treatment as well as the CTD-101 k epoxy resin for vacuum impregnation of the superconducting magnets were identified as the three most valuable parts of the manufacturing process. Subsequently, a method called Technology Competence Leveraging (TCL) was utilized to identify new application fields for each of the selected high technologies and processes. TCL is based on open innovation principles such as crowdsourcing and combines creative and analytical methods to systematically trigger the discovery of new fields of applications in four sequential steps [4]. Following this method, the first step was to identify the key use benefits of these technologies by interviewing experts and actual users. After the identification of the respective use benefits, the goal was to identify as many new application fields as possible by divergent thinking with creativity techniques such as brainstorming and searching for industries that seek to solve similar problems like the ones solved by the technologies. The ideas were clustered and evaluated based on interviews with potential users to assure their applicability. Once a list of probable application fields was established, they were evaluated according to their benefit relevance and strategic fit. These concepts rate the ideas concerning their marketability. While benefit relevance assesses the relevance of the problem to be solved, strategic fit measures the fit of a given idea to the producing company and its resources, capabilities and culture.

For the final step in TCL, the market potential was analyzed by identifying how to derive value from a particular idea, for whom, how to generate income and which key partners are needed to implement the idea. Based on the analysis, the market potential of a new application field could be estimated. An overview of the 21 most promising application fields can be found in Fig. 1 below.

TCL proved to be a very useful tool to identify new market opportunities for technologies involved in manufacturing superconducting magnets. In total, 65 new application fields could be identified using this method. For two of the potential application fields, scrap metal recycling and fruit sorting, a detailed market analysis was conducted in a follow-up project. An overview about the results can be found in the following paragraphs.

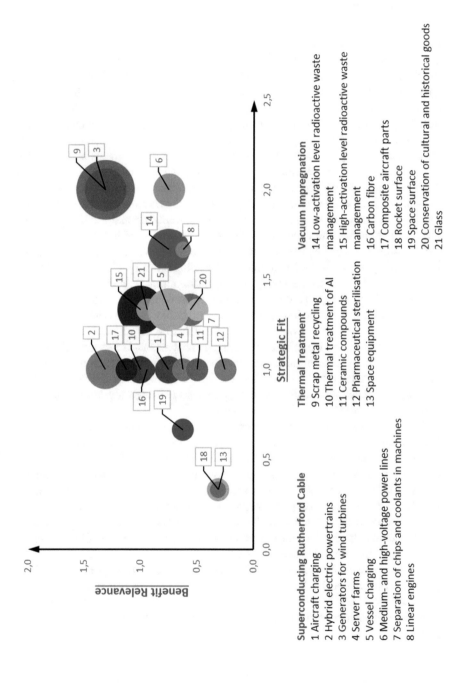

Fig. 1 Evaluation of potential application fields based on their strategic fit and benefit relevance

1 Market Analysis Scrap Metal Recycling Market (Focus: Aluminium)

One potential application field of the furnace used to treat the Rutherford cables is the use in the smelting industry. Compared to the standard furnaces in the industry, it has the advantage that it can reach high temperatures up to 900 °C precisely within an accuracy range of ±3 °C. In comparison, furnaces used in the steel industry have a range of ±10 °C. Furthermore, the heating system generates a homogenous diffusion of heat with internal fans and permits fast temperature changes at a rate of 50 K per hour. By leveraging the protective properties of argon, a noble gas, the furnace additionally minimizes the shrinking of metals due to corrosion (oxidation). This way, the process becomes more efficient and less material is lost, which is important for valuable and expensive materials. These properties are particularly relevant for the metal recycling industry. Utilizing the above-mentioned furnace can reduce the number of steps by combining the splitting up and melting of several metals. Specifically, the technology allows the separation of metals with different melting points. According to interviews conducted with industry experts, the furnace is particularly attractive for the aluminium recycling industry. Currently, 60 Million tons of aluminium are used annually and the consumption is predicted to increase by 18% until 2023. The global demand for this valuable metal is primarily driven by industries such as construction, aviation and automotive, where it has replaced steel due to its high strength to weight ratio. The metal is highly recyclable, with around 75% of all aluminium manufactured, dating as far back as 125 years, still being in use today. The metal is infinitely recyclable without degradation and its recycling process is considered to be closed-looped. Furthermore, the re-melting process of aluminum is much less costly and energy-intensive, using only 5% of the energy needed for the production of new aluminum.

The current recycling process exhibits some challenges such as possible contamination. Impurities in the metal have an impact on its properties and can weaken the metal, rendering it unusable for specific industrial purposes. However, by using the particular furnace for the process, the aluminum could be separated from contaminating metals such as lead and tin, which are primarily responsible for its contamination. Furthermore, during the melting process, a waste by-product called dross develops due to the oxidation of the metal. With argon as a protective gas, the furnace could help to reduce the development of dross, consequently diminishing the loss of aluminum during the recycling process.

The aluminum recycling industry is steadily growing: While primary aluminum production decreased by 8.5% within 5 years, the amount of recycled aluminum increased by 21.6% within the same timeframe until 2015. In Europe alone, this secondary aluminum production is operated by more than 300 aluminum recycling plants.

2 Market Analysis: NMR Technology in the Fruit Sorting Industry

Nuclear magnetic resonance spectroscopy can be utilized to grade and sort fruits and vegetables by their quality. Standard sorting machines in the food industry work with high-frequency cameras that take pictures of vegetables and fruits from different angles. This approach however limits the quality evaluation to external characteristics such as size, color, curvature and visible damages, disregarding potential internal physiological defects. According to industry experts from companies such as VOG, Microtec and Greefa, this is an important concern in the food quality industry and thus, some companies currently employ infrared and UV technology to detect internal defects. However, these technologies are not able to conduct a structural analysis of organic material, determining for example the maturity of the produce, which is an important predictor for shelf life expectancy and storage conditions. In addition to limited conclusiveness about total product quality, many technologies are invasive (thus destroying the fruit or vegetable, time-consuming or too superficial to analyze the fruit core. Nuclear magnetic resonance (NMR) spectroscopy can overcome these challenges by analyzing the chemical composition of the products and detecting parasites, impurities or diseases based on compositional and structural properties.

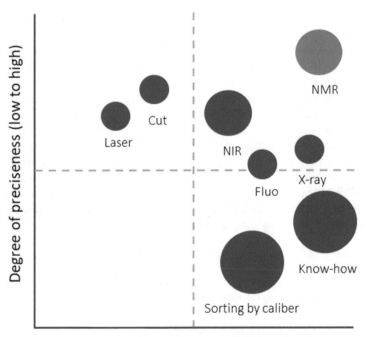

In recent years, the food sorting market has exhibited steady growth, with the average growth rate of 7.24% per year. Factors contributing to the development are higher consumer expectations as well as the demand for higher quality standards within the industry. With its non-invasive properties, NMR-based technology can ensure the quality of organic produce throughout the entire food supply chain and would enable producers to diversify their prices concerning the quality of the product. In 2008, FFV waste of US retail stores due to low quality was estimated at around USD43 Billion (retail price). Since food waste due to spoilage and contamination incurs the most economic costs at the end of the supply chain, an employment of NMR early in the value chain would economically be the most effective option.

3   Learnings

The research projects outlined above revealed promising application fields outside of the typical particle collider domain, which means that benefits derived from the superconducting technology research go beyond building research infrastructures to answer fundamental research questions. While not apparent at first glance, the research about new high technologies comes with profound economic impact on other industries. There are three key takeaways from conducting these projects: First, when evaluating the application potential of high technologies, it is important to consider not only the final product, but also knowledge, processes and technologies involved in manufacturing the product when assessing market potential. Many processes involved in the manufacturing process have great use benefits and thus, high application potential in other industries. Not leveraged, the industrial value of high technologies is systematically underestimated. Second, the identification of new application fields is an iterative process, including many dead ends. It is crucial for the success of such a project to anticipate unexpected turns in finding suitable— and especially valuable—application fields. Finally, in order to produce accurate results, it is essential to interdisciplinary collaboration between scientists and industrial players. Without combining the knowledge from technology experts with market insights from companies and business experts, a diffusion of technological advancements from science to industry is unlikely to be successful.

Acknowledgments The research discussed in this paper is supported by the ASITrain—European Advanced Superconductivity Innovation and Training. The Marie Sklodowska-Curie Action (MSCA) Innovative Training Networks (ITN) receives funding from the European Union's H2020 Framework Programme under grant agreement No. 764879.

References

1. Keinz, P. & Kretzschmar, L. (2019). Impact potentials of EASITrain research on society & industry (Version V1.0). Zenodo. https://zenodo.org/record/3458923#.Xa8VentCRaQ
2. Brzobohaty, L., Habernig, S., Moravec, P., Pably, M., Schürz, T., Kretzschmar, L. & Quach S. (2019). Analysis of potential markets for using technologies in the superconducting magnet value chain (Version 1.0). Zenodo. https://doi.org/10.5281/zenodo.3362855
3. Kretzschmar, L., Mehner, B., Hausberger, M., Ledermüller, F., Mayrhofer, F., Schreiber, D. and Gutleber, J. (2019). Manufacturing process of superconducting magnets: Analysis of manufacturing chain technologies for market-oriented industries (Version 1.0). Zenodo. https://doi.org/10.5281/zenodo.2579834
4. Keinz, P., & Prügl, R. (2010). A user community-based approach to leveraging technological competences: An exploratory case study of a technology start-up from MIT. Creativity and Innovation Management, 19(3), 269–289.

Full Presentation

https://indico.cern.ch/event/727555/contributions/3461288/attachments/1867856/3073626/Linn_Kretzschmar_-Leveraging_the_economic_potential_of_FCCs_technologies_and_processes.pdf

How to Value Public Science Employing Social Big Data?

Maria L. Loureiro and Maria Alló

Contents

1 Introduction

Scientific discoveries can be classified as public goods. Arrow [1] discussed properties of knowledge that make it a public good; highlighting in particular, the fact that it cannot be depleted when shared, and once it is made public others cannot easily be excluded from its use. So, public good is a commodity or service that is provided without profit to all members of a society, either by the government or by a private individual or organization. Thus, a global public good is a public good that goes beyond borders, and CERN scientific output is the perfect example of a global public good.

A crucial issue for the supply of such global public goods is based on their value. When analyzing scientific contributions, it is important to note that there are two kinds of values: use values and non-use values. When talking about science we can identify the use value in the case of patents, licenses, and other market realizations of value. The non-use value is derived from potential market realizations that may be

M. L. Loureiro (✉) · M. Alló
University Santiago de Compostela, A Coruña, Spain
e-mail: maria.loureiro@usc.es

© The Author(s) 2021   93
H. P. Beck, P. Charitos (eds.), *The Economics of Big Science*, Science Policy
Reports, https://doi.org/10.1007/978-3-030-52391-6_13

achieved based on scientific discoveries. The category of non-use values contains those denoted as option values, the bequest values and the existence values. The option value refers to the value given to a resource or service that is of no use today, but maybe extremely valuable in the future. In the case of the bequest value, this denotes the possibility of transmitting knowledge and cultural heritage to future generations; while, the existence value is the willingness to pay to preserve a resource or service because of its mere existence and not necessary because any use or benefit can be derived from it. Therefore, the Total Economic Value (TEV) of science should be considered estimating the sum of the three previous types of values. This a complex estimation and somewhat unknown in the short term. Even in the long run, the TEV can also be difficult to compute in the long term due to the existence of risk and uncertainty.

Thus, going back to the initial question, how can we value public goods? It is important to note that in some cases there are no market prices, therefore one option could be to ask citizens their willingness to pay. But, it is also important to note that the information is crucial and, for example, the participation and perceptions of stakeholders is fundamental. In this respect, we can adopt a direct approach asking citizens about their preferences (where different biases may arise) or an indirect approach where we obtain information from the market to obtain insights about how much we value science.

The aim of this work is to analyze the capacity of big data sources to evaluate the TEV of science; focusing specifically on non-use values. Einav and Levin [2] described how much economic research has evolved in the area of big data and new private datasets, showing a significant amount of questions can be addressed now. As an example, this paper provides an analysis about how we can measure the perceptions that citizens have about CERN through the information collected from the social media Twitter.

2   Data: Twitter Data Collection

Social media can be extremely useful to value global public goods, they provide us unsolicited opinions about current projects and is open data as in the case of Twitter. Specifically, Twitter is a social network that was born in 2006 and the place where we can interact with people around the world. These messages are known as "tweets" and they are limited to 140 characters. According to the data of 2019, Twitter has around 261 million monthly active users worldwide.

Data collection was conducted during October, November, December 2018 and from February–June 2019. This process has been based in the search of tweets through the use of hashtags and keywords. In order to collect the data, requests have been made through the library Tweepy for Python which works with Twitter Streaming API specifying a "keyword" or specific "hashtag". At the same time that tweets were received, another request was made to obtain some additional data regarding the publication of the message. In addition, we were interested in gaining information about the language used. Specifically, the following hashtags were used to download tweets: @CERN, @AtlasExperiment, @LHC News, @CERN-LHC

Live, @ALICE Experiment, @CMS Experiment. On average, during this period, we recorded around 698 tweets in English and 247 in Spanish.

3 Data Visualization: Word Clouds

World clouds are effective ways of showing the most predominant topics in a text. They are frequently used for blogging and micro-posting. In particular, Fig. 1 shows the most relevant topics related to climate change in both countries.

In terms of the predominant words, we observe that Tweeter conversations in Spain are related to "cern", "code", "open", "Microsoft", "software". In the case of English conversations, words related to "cern", "lhc", "collider", "hadron" are the most predominant.

4 Results from Emotion Analysis and Hedonometer

4.1 Analysis of Emotions

In order to analyze our social media dataset from Twitter the lexicon developed by the National Research Center Canada (NRC) is employed; specifically, the dataset EmoLex [3]. The authors focused on eight different emotions: joy, sadness, anger, fear, trust, disgust, surprise and anticipation [4] and the type of sentiment being positive or negative. The lexicon employed contains around 14,000 words and

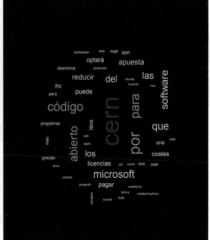

Fig. 1 Word clouds from tweets in English and Spanish

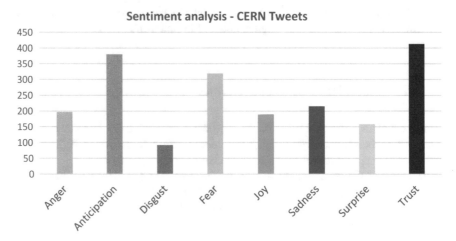

Graph 1 Analysis of emotions: English speakers

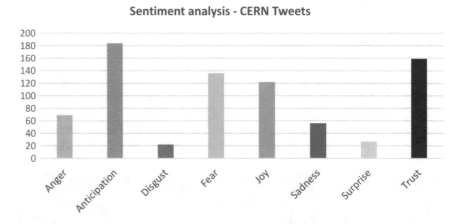

Graph 2 Analysis of emotions: Spanish speakers

25,000 word senses. As indicated, they identified a list of words and phrases and used Amazon's Mechanical Turk to obtain emotions annotations.

The two graphs below show that out of the eight emotions evaluated, the most popular one in both English and Spanish tweets are "trust", and "anticipation" (Graphs 1 and 2).

4.2 *Hedonometer*

In order to analyze the information retrieved from Twitter, the Hedonometer tool has been adopted. Previous studies conducted by Cody et al. [5] have also employed this

Table 1 Hedonometer results

How happy are you from 1 (unhappy) to 9 (very happy)?	
Spanish h_{ave}	5.545564
English h_{ave}	5.353653

technique. Specifically, the Hedonometer technique could be classified within the techniques of Natural Language Processing (NLP). It consists in the analysis of a text through its segmentation into phrases, and phrases, into words. Words are associated with scores of positive and negative feelings, and thus a total score for the sentence obtained, and then by aggregation, for the overall topic. In order to associate each word with a score, specifically, it uses the sentiment scores collected by Kloumann et al. [6] and Dodds et al. [7]. More specifically, the analysis is based on a lexicon or dictionary, which takes as reference, the 10,000 most used words in a long collection of articles from newspapers and books of Google Books as well as lyrics and a large number of generic tweets. These words have been classified according to the values of "happiness" or acceptance that 50 individuals have received from the Amazon Mechanical Turk platform.

The scale of happiness or acceptance has been measured on a Likert scale from 1 to 9, 1 being "least happy" and 9 counting as "most happy", being able to segment the expression in "sadness or rejection" from 1 to 3, "indifference" from 4 to 6 and "happy" or "acceptance" from 7 to 9. The average happiness of users has been calculated following Eq (1):

$$h_{avg}(T) = \frac{\sum_{i=1}^{N} h_{avg}(w_i) f_i}{\sum_{i=1}^{N} f_i} = \sum_{i=1}^{N} h_{avg}(w_i) p_i, \tag{1}$$

Where T refers to a text, f_i is the frequency of the ith word w_i for which we have an estimate of average happiness, $h_{avg}(w_i)$, and $p_i = f_i / \sum_{j=1}^{N} f_j$ is the corresponding normalized frequency. From all these words, we have previously excluded the "stop words" (or words that are necessary for the grammatical construction but that by themselves have no meaning).

Results obtained are presented in Table 1. As it can be seen, the average happiness for both, English and Spanish speakers when talking about CERN is 5.54 and 5.35 respectively.

5 Conclusions

The above results show, accounting for all content published during the period of download, an overall positive feeling towards CERN research activity; and particularly this is reflected by the Spanish speakers. Such positive outlook may support a willingness to pay for CERN scientific services that should be further investigated.

References

1. Arrow, KJ. (1962). The economic implications of learning by doing. The Review of Economic Studies. Oxford Journals. 29 (3): 155–73. doi:https://doi.org/10.2307/2295952.
2. Einav, L. and Levin, J. (2014). Economics in the Age of Big Data. Science, 346, 715-721. https://doi.org/10.1126/science.1243089
3. Mohammad, SM., Turney, PD. (2013). Crowdsourcing a Word–Emotion Association Lexicon, Computational Intelligence, 29 (3), 436-465.
4. Plutchik, R. (1980). A general psychoevolutionary theory of emotion. Emotion: Theory, research, and experience 1, 3–33.
5. Cody, EM; Reagan, AJ; Mitchell, L; Dodds, PS; Danforth, CM. (2015). Climate Change Sentiment on Twitter: An Unsolicited Public Opinion Poll. PLoS ONE 10(8): e0136092. https://doi.org/10.1371/journal.pone.0136092.
6. Kloumann, IM., Danforth, CM., Harris, KD., Bliss, CA., Dodds, PS. (2011) Positivity of the English language. Available at http://arxiv.org/abs/1108.5192 .
7. Dodds, PS; Harris, KD; Kloumann, IM; Bliss, CA; Danforth, CM.(2011). Temporal patterns of happiness and information in a global social network: Hedonometrics and Twitter. PLoS ONE, Vol 6, e26752.

Full Presentation

https://indico.cern.ch/event/727555/contributions/3461289/attachments/1868125/3073452/Valuing_The_Public_Good_dimension_of_Science-final.pdf

R&D, Innovative Collaborations and the Role of Public Policies

Riccardo Crescenzi

Contents

1   Introduction

The generation and diffusion of new knowledge and innovation in national, regional and local economies depends on efforts and investments in Research and Development (R&D). These investments need to be coupled by the presence of appropriate Human Capital and skills in the public and the private sectors in order to absorb and diffuse innovation through the entire economy. Figure 1 summarises the results of recent research on the fundamental drivers of local innovation in Europe [1]. The surface 3D plot shows how innovative output (Z-axis—measured by regional patents) responds to simultaneous changes in R&D (Y-axis—measured by local expenditure in R&D) and Human Capital (X-axis—measured by the presence in the same local economy of individuals with university degrees). For low levels of Human Capital, the patent-R&D relationship is flat, whereas for higher levels of Human Capital intensity, the influence of R&D investments on innovation is positive and increases sharply with a higher level of Human Capital.

Some of the research featured in this chapter has received funding from the European Research Council under the European Union's Horizon 2020 Programme H2020/2014–2020 (Grant Agreement n 639,633-MASSIVE-ERC-2014-STG). All errors and omissions are my own.

R. Crescenzi (✉)
Department of Geography & Environment, London School of Economics, London, UK
e-mail: R.Crescenzi@lse.ac.uk; http://personal.lse.ac.uk/crescenz

H. P. Beck, P. Charitos (eds.), *The Economics of Big Science*, Science Policy Reports, https://doi.org/10.1007/978-3-030-52391-6_14

Fig. 1 The joint effect of
R&D and Human Capital
(HK) on regional patent
intensity K, f(RDr,t,HKr,t).
A 3D surface plot. Source:
Charlot, S. Crescenzi R. and
Musolesi A (2015) *Journal
of Economic Geography*,
Volume 15, Issue
6, November 2015, Pages
1227–1259, https://doi.org/
10.1093/jeg/lbu035. Article
distributed under the terms
of the Creative Commons
Attribution License (http://
creativecommons.org/
licenses/by/4.0/)

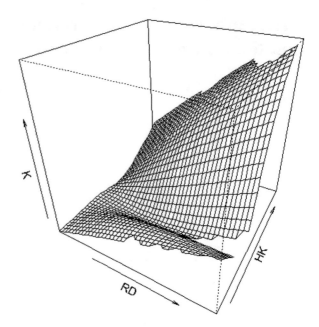

These findings suggest that **R&D investments and facilities do have the potential to boost innovation at the local level.** However, this only happens **where appropriate complementary skills and conditions are available locally** to support knowledge generation and absorption. Investments in R&D can enhance regional innovation only when coupled with a supportive endowment of Human Capital. Both are needed simultaneously to boost innovation, and investing in R&D does not appear to produce a positive effect on innovation for low levels of Human Capital.

2 Balancing the Regional and the Global

In this context, the richer regions of the European Union (EU) benefit from a persistent advantage in terms of the innovation returns to R&D efforts. Conversely, economically disadvantaged regions appear to be in an innovation trap in the sense that a marginal increase in R&D or Human Capital would not increase their ability to innovate. For these regions, investing marginally in such inputs would be wasting money. In particular, the return to R&D expenditure is maximized between 2 and 3% of regional GDP, whereas Human Capital has a positive effect when at least 20% of the regional population has completed tertiary education. The analysis of innovation in EU regions [1] also highlights the presence of shadow effects: high levels of external R&D in neighbouring regions are detrimental for regions with low levels of internal R&D, and the highest joint impact of internal and external R&D is obtained in the correspondence of the highest level of both inputs.

Overall this evidence highlights the risk of 'cathedrals in the desert' scenario, where major R&D investments are concentrated (e.g. because of policy decisions on the location of research infrastructure or localised incentives for private research programmes) in regions that lack the appropriate receptive environment in terms of Human Capital and other systemic conditions. The local mismatch between R&D and skilled labor can persistently hinder innovation and local spillovers.

How can this be avoided? How can public policies facilitate the embeddedness of R&D investments and research infrastructure into local innovation systems? **The key tool is collaboration.**

The romantic notion that a new Nikola Tesla will emerge from the lab with the next AC motor (or an X-ray) increasingly belongs to a bygone era. While in the late 1970s around 75% of EPO patent applications in the United Kingdom (UK) were filed by individual inventors, nowadays that figure is below 15%. More than 80% of all patents are registered to more than one inventor, suggesting that collaboration in research and innovation has become the norm. Teams within the firm or the research centre, but also increasingly complex networks of researchers involving different firms, often in collaboration with universities, public agencies, and research centres drive the world of invention in the early twenty-first century. As Seaborn [2] puts it, "big science [has] eclipsed the garage inventor [. . .] Edison has been superseded by a team of white-coated theoretical physicists".

This fundamental trend towards collaboration in patenting activity is documented in Fig. 2 that plots the share of co-invented patents (i.e. patents filed by two inventors or more) in the United Kingdom since 1978.

While the trend towards the formation of ever-larger research teams and inventor networks has been well documented, we know much less about the factors that drive researchers to collaborate with one another in the first place. How important is geographical proximity and spatial clustering for successful collaborations to happen? What can be done to facilitate local collaborations and spillovers?

Crescenzi et al. [3] have studied empirically the behaviour of 'multiple patent' inventors—i.e. the most prolific and innovative individuals in the economy—showing that being part of the same organisation plays a key role in the formation of co-patenting teams. However, social networks and cognitive proximities are key factors in shaping the selection of team members with a limited direct role of geographical proximity. The role of geographical proximity only emerges as well in interaction with other factors reinforcing their role. This suggests that local collaborations between large research centres and their local environment have the potential to happen but other conditions—to be carefully examined and assessed— need to be in place.

Similar conclusions are reached in [4] looking at the University-Industry collaborations (U–I) collaborations. By looking at the collaborative behaviour of all Italian inventors over the 1978–2007 period, the empirical analysis shows that U–I collaborations are less likely to happen when compared to collaborations involving exclusively university partners of business partners, and suggests that they tend to generate patents of more general applicability in subsequent inventions—measured

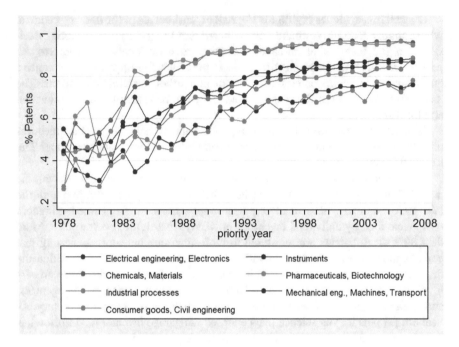

Fig. 2 Co-invented patents by technology field in the United Kingdom, 1978–2007. Source: Crescenzi R., Nathan M., Rodríguez-Pose A. "Do Inventors Talk to Strangers? On Proximity and Collaborative Knowledge Creation", Research Policy, 45(1), 177–194, 2016, doi: https://doi.org/ 10.1016/j.respol.2015.07.003. This is an open access article under the CC BY license (http:// creativecommons.org/licenses/by/4.0/)

by forward-citations. As emphasized by the literature, geographical proximity plays an important role in facilitating all forms of collaboration. At the same time, it works as a possible substitute for institutional proximity, facilitating U–I collaborations. However, the involvement of 'star inventors' on both sides of the collaboration can play an equally important role in 'bridging' universities and industry.

3 Conclusions

Policy-makers have been attracted for a long time by the concept of innovation clusters with the objective of boosting overall regional innovation, development and employment. Public research centres and large research facilities have often been a core part of these local innovation strategies and have allocated substantial public resources to their support and promotion. The rationale behind these policies has been provided by the assumption that geographical clustering would per se support knowledge exchange and innovation. Further analysis on the complementarities between geographical proximity and other forms of proximities is crucial in this regard. An emerging body of evidence seems to increasingly point in the direction of an ancillary role being played by spatial clustering: if other proximity conditions are not simultaneously in place, spatial clustering may—as recent research seems to point out—be of limited utility to innovation. Conversely, public policies might have an important role to play acting as bridges in order to facilitate the development of connections between local teams and those active in the research facility. The presence of star researchers in large research facilities might—for example—be a key factor to facilitate collaboration with local industrial partners offering significant opportunities for technological upgrading.

References

1. Charlot S., Crescenzi R. & Musolesi A. "Econometric Modelling of the Regional Knowledge Production Function in Europe", *Journal of Economic Geography*,15(6), 1227-1259, 2015, doi: https://doi.org/10.1093/jeg/lbu035
2. Seaborn, T., 1979. Talking about the automat. The open channel. Softw. Pat. Inst.(IEEE Comput.) 12, 87–88
3. Crescenzi R., Nathan M., Rodríguez-Pose A. "Do Inventors Talk to Strangers? On Proximity and Collaborative Knowledge Creation", *Research Policy*, 45 (1), 177-194, 2016, doi: https://doi.org/10.1016/j.respol.2015.07.003 [Open Access]
4. Crescenzi R., Filippetti A., Iammorino S., "Academic Inventors: Collaboration and Proximity with Industry", Journal of Technology Transfer, 42(4), 730-762, 2017, doi: https://doi.org/10.1007/s10961-016-9550-z [Open Access]

Full Presentation

https://indico.cern.ch/event/727555/contributions/3461299/attachments/1868119/3072744/Crescenzi_FCC_CERN_June_2019.pdf

Large-Scale Investment in Science: Economic Impact and Social Justice

Massimo Florio

Contents

1   Introduction

Science is not a free lunch. Worldwide, R&D expenditures per year, from basic research to product development by firms, are about USD1.7 trillion (according to UNESCO estimates for 2017). There are perhaps 7.8 million professional researchers globally, around one researcher out of one thousand inhabitants of the planet. In the OECD area, which includes the most developed economies, government R&D spending is worth about USD 315 billion per year and the share of government of the total R&D expenditures is 28%. Hence, citizens support research in two ways: firstly, as consumers by paying a price for goods and services which in turn include in their production costs such expenditures; secondly, by paying taxes which support government R&D expenditures, mostly for basic science.

In this short essay, I discuss two questions: What is the economic impact of basic research? What are the implications for social justice of the interplay between -on one side- government funded science and -on the other side- R&D supported by business? I will argue that the ultimate economic impact of large-scale investment in basic research is often (but not always) positive (i.e. benefits are greater than costs).

M. Florio (✉)
Department of Economics, Management, and Quantitative Methods, University of Milan,
Milan, Italy
e-mail: massimo.florio@unimi.it

© The Author(s) 2021
H. P. Beck, P. Charitos (eds.), *The Economics of Big Science*, Science Policy
Reports, https://doi.org/10.1007/978-3-030-52391-6_15

There is, however, a potential concern for social justice arising from the private appropriation by business of rents arising from knowledge as a public good.

2 Big Science

The main paradigm of knowledge creation for centuries has been based on small scale organizations. Galileo or Newton did not manage large laboratories, and still today there is a considerable role for Little Science: a principal investigator with a small team of postdocs and PhD students. An entirely different organization model has emerged particularly during World War II. The most famous example is the Manhattan Project in the US, that involved thousands of the best physicists and engineers to design and build the first atomic bomb. Another example of this Big Science model is the Apollo Lunar Program, that wanted at the same time to put one man on the Moon, and show to the world that the Cold War could not have been won by the USSR in terms of technological superiority.

There are many other examples: the Nazi atomic bomb program with the involvement of Heisenberg, the French dual nuclear program with both civilian and military missions, the many Soviet 'closed cities' employing tens of thousands of scientists including Nobel laureates, the British secret computer science with Turing, etc.

The main features of the traditional Big Science model are its national scope and defense-related mission, hence a close relation with the military-industrial complex, secrecy, political loyalty of the personnel, top up governance. My claim is that from post-WWII Europe a completely different Big Science model emerged, and CERN is the most important example of it: a large-scale research infrastructure entirely driven by curiosity of a scientific community and explicitly rejecting any connection with the military, based on an international coalition without a main 'owner', adopting open-science style and bottom-up governance. Another example of the new paradigm is the Human Genome Project (1990–2003). The HGP was initially funded by the US Department of Energy (which inherited the Manhattan Project facilities and the study of radioactivity genetic effects), by the US National Institutes of Health, but then involving an international coalition of researchers. In fields as diverse as astronomy, space exploration, materials science, marine biology, and large-scale clinical trials the new research infrastructure (RI) paradigm has successfully changed the landscape of science. Even the contemporary Little Science in fact benefits from access to open data created and distributed by the large RIs, with the crucial possibilities offered by the internet and the World Wide Web (both inventions arising from government funded science, respectively in the US ARPANET agency, and at CERN). The European Bio-Informatics Institute, part of the European Molecular Biology Laboratory (EMBL), a repository of bio-data, is accessed for free every day 70 million times by more than three million unique IP. A PhD student may believe that she is working somewhere in a small laboratory, but in fact she is connected with a global virtual community sharing data, software, and online resources in general.

There is a fundamental difference between the old and the new Big Science paradigm in terms of the opportunity of evaluating their socio-economic impact. Several benefits are similar, for example in the perspective of the economic value of technological progress, but the old Big Science is loaded by military missions which create untreatable problems for social cost-benefit analysis. In fact, there is no sensible way to compare in a global perspective the economic benefits of the Manhattan Project for the US (around 130,000 employees and thousands of companies involved as contractors and suppliers) with the socio-economic effects for Japan of the destruction of Hiroshima and Nagasaki. Apart from any ethical considerations, no cost-benefit analysis of public investment in defense is possible because we do not know how to consider the economic value of war destruction.

When there are no direct military missions, RIs can be evaluated in cost-benefit analysis terms just assuming that the pure knowledge they create will not be harmful. This seemingly modest but crucial assumption paves the way to measurement of RIs' benefits and costs for society using the tools of applied welfare economics. I turn now on this approach.

3 The Net Benefits of Investing in Large-Scale Research Infrastructures

Large-scale investment projects in science are costly. A last generation synchrotron light source has an investment cost of some USD 100 million, and gravitational waves, radio-astronomy, particle physics, nuclear fusion, spallation neutron sources, but also population genomics, or new cancer drugs may require RI budgets in the billion USD scale. In the last 6 years, an interdisciplinary team of the University of Milan and of the Centre for Industrial Studies (CSIL Milano), including economists, statisticians, computer scientists and physicists (particularly Professor Stefano Forte of the UNIMI Department of Physics), has carefully studied the balance of measurable social benefits and costs of the Large Hadron Collider, and the CNAO synchrotron for hadron therapy. The study has been further extended to synchrotron light sources (such as ALBA), distributed RIs for heritage sciences, satellites for Earth Observation, and procurement of the space agencies (particularly of ASI, the Italian Space Agency).

On the cost side, a CSIL team has very recently (2019) provided detailed costing guidelines for the new ESFRI Roadmap (the European Strategy Forum for Research Infrastructures). On the benefits side, in order to avoid double counting and inconsistencies in measurement, our team has suggested a simple intertemporal model of impacts on different social groups: scientists, students and early career researchers, firms involved in procurement, users of products and services embodying technological innovations, users of cultural goods, and finally 'non-users': the general public that includes the taxpayers funding the RIs. The details are explained in Florio [1].

Both costs and benefits must be expressed in a common numeraire, such as Euros or any other currency. In applied welfare economics this is just a measurement convention and does not mean that all the costs (including pollution or traffic congestion) and all the benefits (including visiting an exhibition or a website) actually generate cash transactions. It only means that measurement can be done by an appropriate metric. For example, units of time would also work in principle, and in such case one may estimate whether project costs expressed in time units of standard human effort are greater or smaller than social benefits expressed also in such time units (to see the equivalence just consider that the marginal social value of time is related to what one could earn in money terms even if she doesn't work at all).

The surprising result of the social Cost Benefit Analysis (CBA) of the LHC to 2025 is that the 'side' benefits of the LHC are conservatively estimated to be in excess of its costs with a ratio 1.20; i.e. one Euro of costs returns to society 1.20 benefits. This holds true without considering the unknown value of discovering the Higgs boson or of any other past and future discoveries of the collider (just assuming that such knowledge does no harm). In a new study of the future High-Energy Large Hadron Collider the benefit/cost ratio has been estimated to be 1.8, even greater than the LHC's one [2].

One third of these benefits arise from technological learning for firms involved in the supply chain [3]. These firms, particularly the hi-tech ones, have tested at the LHC new cutting edge technologies, sometimes developed in close contact with scientists and engineers at CERN or in the Collaborations. As a result, firms can take advantage of such knowledge. In economic terms this is an externality and our team has measured very carefully with different econometric methods its impact in terms of additional R&D, patents, productivity, sales, and ultimately profits for CERN suppliers [4].

Another third of the benefits is related to human capital increase for students and early career researchers involved in CERN projects and Collaborations. This is a lifelong effect of skills acquired in a unique scientific and technological environment which translates in a salary premium that we have estimated with different statistical methods based on interviews to current students at CERN, former students now employed elsewhere, and University team leaders. We have also looked at the structure of premium salaries for physicists with different backgrounds (for example with or without specific data analytics skills) [5]. A minor effect for insiders is also related to the narrowly defined value of publications and their influence in the literature, considered just as a product of time and effort (as mentioned without trying to estimate the completely uncertain future economic value of the knowledge embodied in a publication).

Finally, another third of the socio-economic benefits of the LHC arises from use and non-use benefits for the public. The former is related to the implicit willingness to pay for visiting CERN (no admission ticket but there are travel and other costs which point to a value of such visits) or for using cultural products such as websites, traveling exhibitions etc. The latter is the hidden willingness to pay by taxpayers, who de facto fund the LHC. But are they really willing to pay for it?

Two very recent contingent valuation experiments with surveys targeting representative samples of French and Swiss taxpayers respectively, have revealed an estimate of the willingness to pay (WTP) for research in particle physics for the first time worldwide. The surveys were designed and implemented in conformity with the most rigorous standards for such experiments and approved by the Ethics Committee of the University of Milan. In both countries the average respondent, after having inspected some information provided by CERN, has declared her WTP in the form of a tax increase for future investment at CERN. The result is positive and in fact greater than the current implicit tax paid. In France, while around 49 percent of the respondents have WTP = 0 (usually the less educated, low income, and old people), in fact a thin majority of the respondents have a WTP > 0.The overall average is around 4 Euro per year per person, against an actual tax burden around 2.7 Euro. In Switzerland the results are even more positive (with a much lower share of respondents with WTP = 0 and a higher positive WTP), and converging to a much greater value than current tax burden for both a subsample not informed and one informed (during the interview) about how much their taxes pay for CERN [6].

Obviously, the main argument for taking a decision about a large-scale RI must be its scientific case, and one may think solid scientific cases for a RI were measurable social costs—to the best of our knowledge—exceed the predictable benefits. Interestingly, one may also think that to projects where the scientific case is not very strong (perhaps e.g. exploration of Mars with a human mission versus using robots) but the side benefits (for example those related to cultural goods) are strong, a feature well known to NASA managers. It seems in any case that some social cost-benefit analysis is informative as a complement to the scientific cases in order to see a more complete scenario.

While cost-benefit analysis may be able to answer the question of measurable net social benefits at the aggregate level, one may ask a further question: in a planet where inequality and other urgent societal challenges are of major concern, is costly investment in basic science a priority?

4   A Social Justice Perspective

To answer the question one needs to clarify its meaning. There are two different aspects. The first one is whether public money spent in detecting the gravitational waves, the Higgs boson, or other apparently 'un-useful' science should be better spent, for example, for research on cancer or on climate change, or even to immediately provide health care and food for the poor. The second issue is whether the social benefits of investment in science are fairly distributed in terms of equity.

It is important to acknowledge that these are two different questions, and one should not confuse them. The answer to the first question is intrinsically a matter of policy priorities, and there is no serious answer in the perspective of applied welfare economics. One may use the rhetorical argument that science will always produce a benefit for everybody. Looking backwards there may be many good examples.

Scientists do love this argument, particularly when the conversation is about their own project (less so when it is about a rival project or about another field). Unfortunately, the argument is a pretty unscientific one. Clearly, it is not an argument for spending whatever amount of taxpayers' money in whatever research field and forever. Resources in terms of money, personnel, time, energy and materials are not infinite at any given moment. Thus, spending priorities are needed. One may say that building a future particle collider is not a priority, but also one could say that investing in a modern air-carrier of the US or French Navy, which costs several times a LHC, should be delayed until more urgent needs are satisfied. By extension, the same reasoning could apply, however, to most public and private spending as well. This argument hence leads to nowhere. Decision-making on policy priorities and on individual preferences cannot be based on strict social cost-benefit considerations, mainly because of heterogeneity of preferences of individuals and because of radical uncertainty about certain aspects of the distant future. Both economists and scientists should acknowledge that setting policy priorities is part of a political process. One may have opinions about preferred decisions, but it would be difficult to empirically prove how much should be spent by governments in basic science to maximize long-term socio economic benefits. As I mentioned in the Introduction, government funding in the R&D in the OECD area is in the region of USD250 per capita per year. Establishing if it should instead be USD150 or USD350 is an interesting but probably not a better question than the never- ending discussion about how much should be the optimal spending in defense, public universities, health care, or on supporting the fine arts, just to mention some examples.

Having said this, I claim that the different question about the fairness of the current arrangements for government support to science can be investigated in welfare economics terms and in fact may lead to some surprising issues.

Let us consider some examples in biomedical sciences. The Human Genome project had a cumulated cost of around USD3 billion over 12 years entirely supported by taxpayers. It has created new knowledge: we now know that there are around 3 billion base pairs in our genome (by chance the research cost was one dollar per base), and we also know that there are something like 20,000 protein-encoding genes and many more non- protein- encoding ones. Moreover, the knowledge creation process was highly dynamic, new technologies emerged that now allow to analyze a whole human genome at less than one thousand USD and in just 1 h. Databases with millions of such sequenced genomes will be created in the next few years and medicine will be deeply influenced by such knowledge. A social justice issue here arises from the tension between the public good nature of government supported science and the private appropriation of economic benefits.

The HGP data were disseminated according to a form of open science model (the Bermuda conference declaration). In 2013 the US Supreme Court ruled that human genes cannot be patented. However, several hundreds of new biotech products and several thousand patents were derived by the HGP knowledge. Huge capital gains and profits were created in the health industry after the most risky and long-term basic research was paid by the taxpayers. Taxation of capital income is notoriously lower than taxation of labor everywhere in the world, and one may easily see that there has been a transfer of wealth from the average taxpayer to the average investor in the biotech industry, who de facto enjoyed a considerable discount on R&D investment costs. In a counterfactual scenario, one may have designed more effective ways to defend the public good nature of knowledge of the human genome. A perhaps even more revealing case, of a different nature, is the role played for R&D on drugs by the National Institutes of Health (which is worldwide the most important RI -or set of RIs- for biomedical sciences). According to a recent study [7], the R&D of all 210 drugs later approved in the US received NIH funds at the average level of USD840 million per each new drug (and some of such drugs have been supported also by other funding agencies as well). Statistics on internal R&D by private companies is opaque to say the least. Some studies claim that they need to spend USD1.4 billion on R&D per each drug of their own funds. Other studies [8] suggest, however, that this figure is exaggerated for regulatory reasons, and the true value could be around USD648 million for each recent cancer drug. In any case, these estimates of R&D spending by firms imply that the government is a de facto a major funding partner for pharma companies with a 45–55% share of R&D. Compounded with the extremely high prices of new drugs, this leads to high margins of the pharmaceutical industry (after-tax margins around 24% after capitalizing R&D and leases, according to Prof. A. Damodoran of New York University, cited by [9]). It is apparent that citizens are paying twice the bill: firstly, as taxpayers supporting the riskiest part of the research, later as patients directly or indirectly paying the price of drugs. Moreover, patients also donate their data to firms supporting large scale clinical trials. These two examples of asymmetric effects of government funding of science are part of a much wider panorama. As mentioned, CERN has been instrumental to many innovations, including the World Wide Web, and ARPANET to the internet. A large part of the innovations related to computer science and the digital economy can be traced back to government supported research [10]. But it is also apparent that a considerable component of social inequality of our days is related to the fact that investors in the big digital knowledge-based companies (now ranked top of the world by market value) have accumulated huge wealth because they have been able to privately appropriate the economic benefits of scientific and technological knowledge released for free by government- funded research institutions.

There is a paradox in the current research infrastructure model: the more it creates path-breaking knowledge as a public good with the support of taxpayers, the more monopolistic or oligopolistic private companies are able to prosper and extract rents from consumers. Modern big science critically contributes to economic growth and prosperity, but we need to think again how government funded research can contribute to social justice as well.

References

1. Florio M. (2019). *Investing in Science. Social Cost-Benefit Analysis of Research Infrastructures*. The MIT Press.
2. Bastianin A., and Florio M. (2018). *Social Cost-Benefit Analysis of HL-LHC*. https://cds.cern.ch/record/2319300/files/CERN-ACC-2018-0014.pdf
3. Castelnovo P., Florio M., Forte S., Rossi L., and Sirtori E. (2018). "The economic impact of technological procurement for large-scale research infrastructures: Evidence from the Large Hadron Collider at CERN." *Research Policy, 47*(9), 1853-1867.
4. Florio M., Giffoni F., Giunta A., and Sirtori E. (2018). "Big science, learning, and innovation: evidence from CERN procurement". *Industrial and Corporate Change, 27*(5), 915-936.
5. Camporesi T., Catalano G., Florio M., and Giffoni F. (2017). "Experiential learning in high energy physics: a survey of students at the LHC." *European Journal of Physics, 38*(2), 025703
6. Florio M., and Giffoni F. (2018). *Scientific Research at CERN as a Public Good: A Survey to French Citizens*. http://cds.cern.ch/record/2635861/files/CERN-ACC-2018-0024.pdf
7. Cleary E. G., Beierlein J. M., Khanuja N. S., McNamee L. M., and Ledley F. D. (2018). "Contribution of NIH funding to new drug approvals 2010–2016." *Proceedings of the National Academy of Sciences, 115*(10), 2329-2334.
8. Prasad V., and Mailankody S. (2017). "Research and Development Spending to Bring a Single Cancer Drug to Market and Revenues after Approval" *JAMA internal medicine, 177*(11), 1569-1575.
9. The Economist (2019). "Pharmaceuticals. Profit warning", June 22, 53.
10. Mazzucato M. (2015). *The Entrepreneurial State: Debunking Public vs. Private Sector Myths*. Anthem Press.

Investing in Fundamental Research: For Whom? A Philosopher's Perspective

Michela Massimi

Contents

1   The Importance of Investing in Fundamental Research

In uncertain economic times and a volatile political landscape, the question of our panel discussion might strike as otiose. Faced with major and pressing challenges—from climate change to biomedical research, from cybersecurity to agrotechnology, just to mention some examples—our society seems already to be struggling in meeting research targets that affect millions of people around the globe. Why should we care about fundamental research? Tight national budgets often force hard choices about which kind of investments should be prioritised. And if cuts have to be made, investments in fundamental research tend to be the first ones in the line.

In what follows, I make some brief remarks about the importance of fundamental research for society—be it fundamental research in particle physics, cosmology, or other areas. I will make some specific comments about how I see fundamental research contributing to human cultural flourishing and conclude with some reflections about scientific progress in pursuing fundamental research.

A ground-clearing remark is first in order, though. Fundamental research is a misnomer, and on occasions, an unfortunate one too. For it suggests that research comes in two varieties: the fundamental and the non-fundamental. Or, the abstract and the applied, to use another dichotomy. Thus, posing the question about the value of fundamental research inevitably invites a wave of scepticism if the underlying assumption is that a choice is forced upon us between the abstract and the applied.

M. Massimi (✉)
School of Philosophy, Psychology and Language Sciences, University of Edinburgh, Edinburgh, UK
e-mail: michela.massimi@ed.ac.uk

© The Author(s) 2021
H. P. Beck, P. Charitos (eds.), *The Economics of Big Science*, Science Policy Reports, https://doi.org/10.1007/978-3-030-52391-6_16

But reality is a lot more nuanced that this dichotomy might suggest. For the boundaries between research into the foundations of a particular field feeds seamlessly into technological innovation. And technological innovation in turn informs directions of research in fundamental areas. High energy physics is a case in point. It is for example well-known that innovations such as the World Wide Web were developed by scientists working at CERN. Technology such as the PET scan were spearheaded at CERN while developing new technology for particle physics. And more recently research on proton therapy has been carried out at CERN in the fight against cancer.

But leaving here aside these general considerations, there are other reasons as to why investing in fundamental research is important. These general considerations have got to do with what philosophers often describe as a duty of care that we have towards our 'later' selves, or the next unborn generations.

Imagine someone's life as a slow motion cartoon where if the speed is sufficiently slow, you can almost see the individual snapshots the carton consists in. Each of these snapshots is indexed at a particular time t_1, t_2, t_3,... t_n. The question arises as to whether choices made at a particular time, say t_3 should be done with an eye to benefitting the subject *at the time*, or with an eye also to benefitting the later self at time t_n. For example, one might enjoy smoking a cigarette at time t_3 and be careless about the long-term consequences of her/his choices at t_3. Or one might become mindful of the long-term health risks that such action might engender at t_n and decide at t_3 to cut on smoking if at t_3 one cares enough about one's later self and wellbeing at t_n.

Philosophical theories of personal identity and intergenerational justice depend on how we might be inclined to answer temporally-indexed questions of this nature. If we see ourselves as part of a continuum spectrum, where actions and decisions taken at any particular time are bound to affect the well-being and flourishing of our later selves at later times, then I think an easy answer is available to our overarching question, i.e. "Investing in fundamental research—for whom?". The simple and straightforward answer is: "For *our later selves*".

Investing in fundamental research is indeed for our later selves, and for the next generations. More broadly, it is for *humankind*. We do not invest in fundamental research for the sake of some immediate economic return that someone somewhere is directly going to benefit from. But for advancing scientific knowledge, for exploring uncharted territories, and for making progress in our collective understanding of the natural world we live in.

Thus, in a way, this is a very long-term and open-ended kind of investment. And in assessing the value of this kind of investment, in the light of the aforementioned duty of care towards our later selves, my inclination is to warn against a principle that economists tend to use all the time: the *principle of discounting the future*. The economic principle that a dollar today is more valuable than a dollar tomorrow is a bad principle for assessing the *intrinsic value* (as opposed to the cost-effects economic value) of scientific research in general, I think. This is the case no matter whether the research in question is about climate science, cancer research, or fundamental physics. We have a duty towards future unborn generations to study and understand the effects of anthropogenic climate change, to study and monitor the long-term risks of smoking cigarettes, as much as we owe to them a better understanding of the natural world we live in.

One reasonable concern one might have (and some people do indeed have) about investing in fundamental research is not however (or is not only) about the economic costs attached to it, or the long-term risky returns of the investments. In a way, investments of this nature invite us to pause and think about the progress made in the specific area, and to ask ourselves the extent to which further funding is warranted. We live in a society that is obsessed with metrics and ways of measuring progress towards targets. And when it comes to fundamental research in pretty much open-ended areas—be they particle physics or cosmology for example—even setting targets to achieve might prove very difficult.

Because of its very nature, fundamental research tends to be exploratory and open-ended, with targets that ought to be realistic but at the same time are as revisable and as open-ended as the field of inquiry itself. For it would be naïve at best to expect fundamental research to be railroaded in some predefined way towards pre-given fixed targets. Some of the most important breakthroughs might happen in unexpected ways, while some pre-given targets might prove unachievable, after all.

Here I think we need to be careful and not fall prey to a very common but in my view short-sighted view of how to assess scientific success and progress when it comes to fundamental research. There is a deeply instilled tendency to measure scientific progress and success in any given area in terms of numbers of scientific discoveries and the potential for discovery; by setting milestones and monitoring how well we have marched to achieve those. Think of it. It is a pervasive view that science progresses by discovering new things, or by inventing new tools or enabling new technological innovation. We discover a new particle; patent a new vaccine; unveil a new phenomenon.

Unsurprisingly, a sense of frustration tends to accompany areas of science where the notion of progress does not necessarily conform to this received and somewhat intuitively satisfying notion. High energy physics is one of those areas where frustration might occasionally become tangible among practitioners and the general public too. Will we find new physics beyond the Standard Model? Will the new generation of colliders deliver on the promise of finding new phenomena and shed light on various unsolved puzzles? Given the open-ended nature of fundamental research how to precisely answer these questions is far from trivial and obvious.

But here is my positive take-home message. Even in the worst-case scenario, even if we were not able to discover new particles, new phenomena, new physics beyond the Standard Model as particle physicists hope for (and have reasons for hoping for), we would still have made progress. Because scientific progress is not necessarily (or exclusively) discovery-driven. Progress in fundamental research is not just about finding out what is actual, but also (and equally importantly) is about *delimiting the space of what is possible*. Progress in HEP often means being able to rule out live options, carve out the space of what is reasonable to expect, or of what we believe to be possible to the best of our knowledge. That is what physicists at the LHC, CERN, have been doing over decades. They have fixed more rigorous constraints to rule out possible candidates for Beyond Standard Model physics. They have run high-precision measurements to refine our understanding of the Standard Model. They have clarified where the boundaries of the current space of possibilities lie. This is progress in physics and in science, more in general.

Consider an analogy with cancer research. Yes, we have not yet found as of today (July 2020) a final and definitive cure for cancer. But should we then conclude that there has not been progress made in fighting cancer? Of course, there has been huge progress made in better understanding the mechanisms of carcinogenesis and the specific details of particular kinds of cancers. Such understanding has advanced our ability to have targeted treatments and to improve the overall prognosis for millions of people worldwide.

Along similar lines, we might not have found as I write this piece (in July 2020) the key to various outstanding mysteries about the low mass of the Higgs boson and the nature of dark matter, among many others. But we do have made huge progress in better understanding the gaps in the existing Standard Model, in conceiving possible theoretical solutions for them, ruling them out too, and in setting more rigorous constraints on what might be possible on the basis of the available evidence. This is progress. Indeed this is the type of progress that warrants further funding investment even if the field is exploratory and open-ended.

The real question then becomes the following. How can we make sure that such investments are responsible and engage with local communities? How can we design relevant infrastructures that can be used and reused? And can we supervise and monitor the training and education of the next generations of scientists and make sure they gain a set of skills that are transferable and with some clear pathways for a wide range of applications? These are pressing and open questions to which I hope philosophers of science will contribute more and more in this ongoing dialogue and engagement with physicists.

Acknowledgments This article feeds into a larger project that has received funding from the European Research Council (ERC) under the European Union's Horizon 2020 research and innovation programme (grant agreement European Consolidator Grant H2020-ERC-2014-CoG 647272 *Perspectival Realism. Science, Knowledge, and Truth from a Human Vantage Point*).

Investing in Fundamental Research: Evaluation of the Benefits that the UK Has Derived from CERN

Philip Amison and Neil Brown

Contents

1 Introduction

Investing in fundamental research enables us to push the frontiers of knowledge. However, since we don't have access to unlimited resources we have to make choices about which areas to invest in. It is often easier to place a value on the costs of an investment than the benefits. Often, the benefits of funding a piece of research cannot be known in advance of making the investment, plus it may take many years for those benefits to be realised. Even in cases where we are looking back—trying to place a value on a past or ongoing investment in research—it is not always straightforward to identify, attribute and quantify the benefits and it is typically harder still to place a monetary value upon them.

Even if we think we understand the pros and cons, or the costs and benefits, of investment in research it is important to communicate and consider that understanding with a wider public, including the ultimate funders of public research, the

P. Amison (✉)
Science and Technology Facilities Council, UK Research and Innovation, Swindon, UK
e-mail: philip.amison@stfc.ukri.org

N. Brown
Technopolis Group, Brussels, Belgium

© The Author(s) 2021
H. P. Beck, P. Charitos (eds.), *The Economics of Big Science*, Science Policy Reports, https://doi.org/10.1007/978-3-030-52391-6_17

taxpayer. Articulating the benefits from research is also important if we are to continue to inspire society more generally to engage with science and technology and encourage the next generation to study and work in this area.

To help answer questions such as these, in 2018 the Science and Technology Facilities Council (STFC)[1] commissioned Technopolis[2] to undertake an evaluation of the benefits that the UK has derived from CERN. The aim was to capture, demonstrate and measure the range of scientific, economic and social impacts emerging over the past decade, considering both direct UK involvement and use, as well as any wider influences of CERN on the UK. The study drew on multiple sources of evidence, including desk research, surveys, interviews, case studies and bibliometrics, to explore the various impacts of CERN on the UK.

The following sections of this paper summarise the main findings from the study. Further information about the study can be found in the main report, which will be available soon via the UKRI web pages.

2   The UK's Involvement in CERN

The UK is one of CERN's founding members and has been centrally involved throughout its history. It currently contributes around £150 m per year to the CERN budget (16% of Member State subscriptions in 2019), which covers the building, operation and maintenance of the infrastructure, plus the governance and administration of CERN. The construction, maintenance, upgrade and operation of the experimental programme and the computing infrastructure is then mainly supported through funding from agencies of participating countries. For the UK this funding—as well as direct involvement in construction, maintenance, upgrades and operation—is mainly provided through STFC, enabling UK-based researchers to participate in the experimental programmes hosted by CERN.[3] Through its subscription and broader participation the UK secures a number of benefits. These include:

- Access for UK physicists/engineers to key research infrastructure and collaboration networks;
- The opportunity for UK companies to bid for contracts, including those requiring a high intellectual and technical capacity, those that are non-technical but require high levels of expertise (e.g. financial services) and contracts for more standard supplies and services (e.g. cabling); and

---

[1]The Science and Technology Facilities Council (STFC) is part of UK Research and Innovation; www.ukri.org

[2]Technopolis group, United Kingdom, 3 Pavilion Buildings, Brighton BN1 1EE, T + 441,273 204,320; E neil.brown@technopolis-group.com; www.technopolis-group.com

[3]Other parts of UKRI and other funders also award funds relating to CERN, but on a smaller scale.

- Training and work opportunities, e.g. long-term attachments for students from UK universities; visit programmes for UK schools and teachers; and apprenticeships, secondments and fellow's schemes.

It is important to note that CERN is not a user facility in the way that many other research infrastructures are (e.g. the ILL or ESRF). Rather, the UK is a partner in the co-development of the CERN facility and its programmes of work. CERN is also the UK's national laboratory for particle physics and the UK has played a key role in its strategy and development, while UK personnel have been involved in all the major experiments and discoveries. When we talk about CERN's activities and achievements, therefore, these are really the results of cross-country collaborative efforts and endeavours.

3   Main Pathways to UK Benefits and Impact

CERN's strands of activity and engagement are multi-faceted, with a wide range of types of benefits and impact, flowing through a series of interrelated pathways. The study defined and demonstrated 12 main areas of impact that flow to the UK, which are organised under the broad impact areas of research, innovation, skills and science diplomacy. Figure 1, below, summarises each of these impact pathways. The highlights from each of these areas are summarised in the remaining sections of this paper.

4   Benefits and Impacts Relating to World-Class Research

CERN research makes important advances in particle physics, including a series of landmark discoveries such as the Higgs boson—the last missing piece of the Standard Model of particle physics, hypothesised more than 40 years earlier. Other major breakthroughs made at CERN include the discovery of weak neutral currents (1970s) and electroweak (W and Z) gauge bosons (1980s), measurement of the number of lepton generations (1990s), observation of CP violation in charm quarks (2019) and the (to-date) null result showing the lack of supersymmetry. These advances support further scientific progress and offer the potential for very significant wider societal impacts in the longer term. UK scientists also build on this research to support their further progress and achievements. In the past 10 years alone, over 20,000 UK scientific papers have cited CERN articles, and this includes many of the UK's most influential physics papers (25% of these UK papers are among the 10% most cited in their field globally).

Pooled investments by Member States have enabled facilities that would be beyond the reach of any country alone, providing access and opportunities at world leading facilities for UK researchers at a fraction of the overall costs. This

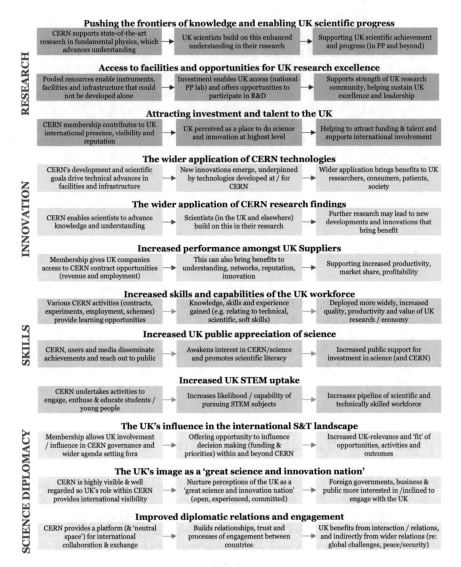

Fig. 1 Main pathways to UK benefits and impacts from CERN. Source: Evaluation of the Benefits that the UK has derived from CERN, Technopolis Group, 2019

includes access to unique technologies and capabilities, international collaborations and networks, frontier science and experiments, the latest theories and developments, new methods, training and learning opportunities. The UK science and engineering community is taking up such opportunities on a significant scale, with over 1000 researchers from 30 UK organisations currently using CERN (third highest amongst Member States).

CERN opportunities support the strength of the UK research community, helping to sustain the UK as a world leading research nation. UK personnel have been involved in all of the major experiments and discoveries at CERN, with many UK researchers holding key positions. CERN-based publications have also significantly pulled the UK's citation metrics upwards, demonstrating the high-quality research enabled. In addition, CERN contributes to the UK's international presence, visibility and reputation, which plays out through the attraction of funding, talent and other forms of recognition for the UK.

5 Benefits and Impacts Relating to World-Class Innovation

CERN's scientific breakthroughs have required major advances in technologies, which have then found wider application, across research and industry. In several notable cases (e.g. the World Wide Web, detectors for PET scanners, touchscreens, GRID computing), CERN has provided the platform for a major new technology that has come into general use and had a transformative effect, bringing economic and societal benefits to the UK and the rest of the world. Other examples of innovations emerging from CERN include the HTTP protocol, next-generation (hadron beam) radiotherapy, radiation-hardened robotics for nuclear decommissioning, fibre optic sensors to help manage water shortages and various software tools and advances in machine learning, pattern recognition and big data analyses (amongst many others). Forthcoming upgrades to the facility will require further technological innovation, which in time will no doubt also find wider uptake and application beyond CERN and particle physics.

CERN membership also gives UK companies access to a steady stream of contract opportunities, with around 500 UK firms having sold goods and services to CERN in the past decade, bringing in an additional £183 m in revenue and supporting employment (all figures in 2018 prices, unless stated). In addition, at least £33 m was awarded to UK firms for CERN experiments (organised by collaborating countries) and by the CERN Pension Fund. These contracts have been won by a wide range of UK firms, from small precision engineering companies, through to global technology firms and pension fund managers.

These UK suppliers also realise wider benefits, beyond the value of the contracts themselves, for instance through the development of innovative technologies or access to new market opportunities. CERN contracts also bestow a degree of prestige on suppliers that is not easily replicated elsewhere, which aids new sales. Half of UK suppliers reported that past CERN contracts had resulted in an increase in other sales income, and the study estimates that a further £1 billion in turnover and £110 m in profit has been supported amongst UK suppliers in the past decade, on top of the direct income received through contracts.

6   Benefits and Impacts Relating to World-Class Skills

There is significant uptake of CERN training opportunities in the UK. In the past decade, around 1000 individuals have participated across the various formal schemes that CERN offers, receiving (free) training worth more than £4.9 m. Even more acquire skills and knowledge 'on the job', including each year around 1000 researchers, 300 CERN staff, 40 fellows and hundreds of individuals at UK suppliers. Through these interactions, the UK workforce gains knowledge and skills across a variety of areas (technical, scientific, digital, problem solving) through an experience that is considered near-unique. Young UK researchers who have engaged with CERN are estimated to earn 12% more across their careers as a result (with an extra £489 m in additional wages realised in the past 10 years alone).

The knowledge and skills gained via CERN are also deployed more widely in the UK economy. Students, researchers and staff move to various roles (analysts, scientists, engineers, developers, management) across a variety of sectors (IT and software, engineering, manufacturing, financial services, health) in the public, private and third sectors. Their capabilities are in great demand, with shortages of STEM skills in general costing UK firms £1.5 billion a year in recruitment, temporary staff and additional training. CERN, researchers and the media also disseminate and reach out to the wider UK public. Each year (on average) from the UK: 12,000 school students and other members of the public visit CERN in person; 220,000 visit CERN's website; and 40,000 interact with its social media. There are also 2000 mentions of CERN in the UK media each year, plus various TV/Radio broadcasts. CERN helps to increase the UK public's appreciation of science, awakening interest in CERN, the science that it supports and the benefits of this work. This helps promote scientific literacy and in the development of a culture valuing science. The results of a separate study (Florio, 2018) suggest that the UK public would be willing to pay (through taxation) around £1.2 billion for CERN over a decade—more than the UK's actual contributions.

CERN also undertakes activities specifically aimed at engaging, enthusing and educating young people. The UK has the most teachers attending the CERN National Teacher Programme (over 1000 in the past decade), who go on to teach an estimated 175,000 school students with context from CERN within 3 months of the visit. As mentioned above, thousands of UK pupils also visit CERN each year, increasing the likelihood and capability of young people pursuing STEM subjects at A-level and university.

7   Benefits and Impacts Relating to Science Diplomacy

The UK is actively involved at all levels of CERN governance, providing UK ministries, funding agencies, and the wider UK science base with an important platform for international engagement, leadership and agenda-setting. CERN also provides a platform for the UK to engage more widely in global initiatives and international networks. CERN is highly visible and well regarded internationally, which spills over to favourable perceptions of its members and greater engagement (in science, technology and beyond).

CERN has been instrumental in science diplomacy, with a constitutionally-defined policy of openness and a commitment to provide a neutral space for global collaboration, the importance of which was recognised when CERN was granted observer status at the UN General Assembly. CERN also actively seeks to establish links with and promote research by countries across the globe, facilitating cooperation among the scientists and policy makers of many countries that are experiencing strained relations at the political level. CERN is an example of science for peace that has inspired several other major cooperative initiatives. For example, the SESAME synchrotron light source showcases the critical role that CERN has played in fostering cooperation across political, religious and cultural divides in the Middle East.

CERN has also nurtured the global physics community through the development of young researchers (particularly in less well-endowed scientific communities) and via collaborations. UK scientists have been at the centre of this outreach work that has resulted in many countries—from Argentina to South Africa—engaging with CERN experiments and investing in their own national researchers and facilities, with improved local capabilities and wider international research collaborations resulting.

8   Conclusion

CERN addresses fundamental questions about the Universe at a facility of unprecedented scale. It is the world's largest particle physics laboratory, making available complex, purpose-built particle accelerators and detectors, as well as computing technology, for its global research community. Its strands of activity and engagement are multi-faceted—from fundamental research to discover new particles and forces, through to school visits that inspire tomorrow's scientists—and this study has provided the most thorough attempt yet to capture and measure the full range of benefits that flow from CERN, covering impacts relating to research, innovation, skills and science diplomacy.

Monetising such impacts is challenging, however, as the laboratory's contributions to society unfold over decades, with its advances in knowledge and technology helping to underpin manifold social and economic benefits that occur in many different places and in a somewhat unpredictable fashion. Moreover, the realisation of these benefits invariably depends, in part at least, on many accompanying developments and wider events. In addition to these CERN-derived knowledge spillovers, there are also undeniable benefits of being a member and 'sitting at the table' where decisions are being made today about the future of particle physics 20–30 years hence.

Evaluators are gradually developing tools to overcome the particular methodo-logical challenges of tracing and measuring the impact of public investment in Big Science, and Technopolis' study for STFC used a range of different state-of-the-art approaches to capture and monetise some of the most important impacts attributable to CERN, demonstrating over £1 billion-worth of benefits for the UK in the period 2009–2018. This is a lower bound estimate of overall impact and doesn't, for instance, include wider technology spillovers (e.g. the World Wide Web), which if monetised would be substantial. Even so, it already largely justifies the cost of UK investment in the facility, and provides a good first indication of the scale of potential benefits that flow from our investment in such large scale research infrastructures.

Fundamental Science Drives Innovation

Carsten P. Welsch

Contents

Listening to the words "fundamental science" many people often think that this concept refers to something distant from their daily lives; a fact that already poses a challenge when thinking who and how profits from investments in these fields.

It turns out though that fundamental research is interwoven with our everyday life. Today, there are more than 50,000 particle accelerators in the world [1], ranging from the linear accelerators used for cancer therapy in modern hospitals to the giant 'atom-smashers' at international particle physics laboratories used to unlock the secrets of creation. For many decades these scientific instruments have formed one of the main pillars of modern research across scientific disciplines and countries.

Many of today's most advanced research infrastructures rely on the use of particle accelerators. This includes for example synchrotron-based light sources and Free Electron Lasers, high energy accelerators for particle physics experiments, such as the Large Hadron Collider (LHC), high intensity hadron accelerators for the generation of exotic beams and spallation sources, as well as much smaller accelerator facilities where cooled beams of specific (exotic) particles are provided for precision experiments and fundamental studies.

Much less known is the fact that particle accelerators are also very important for many commercial applications, such as for example medical applications, where they are used for the provision of radioactive isotopes, x-ray or particle beam

C. P. Welsch (✉)
Physics Department University of Liverpool and Cockcroft Institute, Liverpool, UK
e-mail: C.P.Welsch@liverpool.ac.uk

© The Author(s) 2021
H. P. Beck, P. Charitos (eds.), *The Economics of Big Science*, Science Policy
Reports, https://doi.org/10.1007/978-3-030-52391-6_18

therapy. Furthermore, they are widely used for material studies and treatment, lithography, or security applications, such as scanners at airports or cargo stations.

This article first presents how accelerators enable scientific discoveries, before discussing some of the applications that now benefit our society on a daily basis. It closes by showing the unique training opportunities offered by this interdisciplinary field.

1   Fundamentals Science: For the Advancement of Humankind

Curiosity-driven fundamental research has driven revolutionary transformations of society, such as the rapid growth of computer-based intelligence and the discovery of the genetic basis of life. Albert Einstein's famous theory of relativity is now used every day as part of the Global Positioning System (GPS) and built into mobile phones and car navigation systems.

Fundamental research not only radically alters our understanding of the world around us, it also leads to new tools and techniques that transform society, such as the World Wide Web, originally developed by particle physicists at CERN to foster scientific collaboration. Cutting edge research requires the sharpest minds and needs them to work together on some of the hardest challenges. The outcomes of these collaborative studies then often have an earth-shattering impact on our everyday lives.

The path from exploratory fundamental research to society applications is, however, not direct nor is it predictable. Sometimes, new technologies enable even more fundamental discoveries, e.g. quantum mechanics, and these in turn are the basics for applications such as quantum computing which has huge potential to revolutionize the way we use computers altogether.

To use the full potential of human intellect and innovation, it is important to find a good balance between finding solutions to short-term problems and at the same time enabling real transformational studies that usually come from serendipitous discoveries [2].

Unfortunately, decreasing funding for research, combined with economic and political uncertainty, has led to a focus on short-term goals that often help address current problems. However, these risks miss the huge transformational discoveries that historically, almost always, arise from fundamental research.

Particle accelerators have been one of the driving forces behind scientific discoveries and, in turn, ground-breaking innovations. The need for higher-energy beams for fundamental research as compared to those found from natural radioactive sources has been the major motivation for advances in particle accelerators.

The LHC is currently the world's largest particle collider allowing the global particle physics community to explore nature at its most fundamental scales. It is hosted in a 27 km circumference tunnel beneath the France-Switzerland borders in the Geneva area. The LHC was built thanks to the collaboration of about 10,000 scientists and hundreds of universities and laboratories from more than 100 countries

Already in 1927, Lord Ernest Rutherford demanded a "copious supply" of projectiles with higher energies, as natural α and β particles would provide. When he opened his High Tension Laboratory, he stated that "we require an apparatus to give us a potential of the order of 10 million volts which can be safely accommodated in a reasonably sized room and operated by a few kilowatts of power. We require an exhausted tube capable of withstanding this voltage." John Cockcroft and Ernest Walton picked up this specific challenge and invented the high-voltage generator that is now named after them.

Almost 100 years later, accelerators are still at the core of scientific discovery and enable research groups from around the world to work together on some of the biggest scientific challenges. The LHC has enabled the discovery of the Higgs Boson—the last missing piece in the Standard Model of Particle Physics—and scientists are currently planning to build an even better microscope to understand the building bricks of our universe even better. This will have to be done in a truly global effort, where generations of researchers work across disciplinary and country borders. *"Science knows no borders"*, said former CERN Director General Rolf Dieter Heuer in a recently produced film about the Future Circular Collider study [3].

2   Accelerating Society

Curiosity-driven research requires and drives innovation in the research techniques and technologies underpinning scientific studies. High(er) power magnets required for controlling the movement of ever-higher energy particle beams for example, readily find application in MRI scanners in hospitals or can help find honey launderers [4]. Innovations resulting from fundamental science studies usually also find application in other areas that benefit society in various ways. Particle accelerators are no exception to this.

Passengers at London's Heathrow Airport got some good news recently when it was announced that—thanks to the airport's new computerized tomography (CT) scanners—they will soon be able to stop separating out the liquids and gels in their hand luggage as they go through security. The new scanners produce high-resolution, three-dimensional X-ray images in real time, making it easier to detect explosives quickly, without the need for a separate screening process. This has been achieved, in part, by improvements to the accelerators that provide the electron beams for the scanners [5] and the image processing techniques. A clear example of progress that was made possible through advancement in technologies and tools which originally targeted fundamental research.

Another example of technology transfer in accelerator science relates to cancer treatment using proton and ion beams. This technique takes advantages from the so-called "Bragg peak"—the fact that protons when going through matter (i.e. a patient's body) do not pass all the way through the body. Instead, they stop sharply at a specific depth determined by their energy. By modulating the beam's energy and direction, one can deliver a specific treatment dose over a 3D tumour volume while sparing healthy surrounding tissue. An international R&D effort has focused on the development of novel beam and patient imaging techniques, studies into enhanced biological and physical simulation models using Monte Carlo codes, and research into facility design and optimization to ensure optimum patient treatment along with maximum efficiency [6]. Collaborative research within the Optimization of Medical Accelerators (OMA) project for example has helped improve cancer treatment using ion beams. Future studies will now look into making this technology more accessible and more abundant in number. Scientists and engineers will be working hard on reducing the entry costs for users in medical imaging, cancer treatment, security and materials science [2].

3   Training the Next Generation

Cutting-edge fundamental science requires our best scientific minds to calculate, observe, and invent together in a way that leads to the next innovation. This attracts scientists and engineers at an early stage and allows for high quality training that is increasingly cross-sector, interdisciplinary and international. International links, research and knowledge exchange are all aspects that help enhance the education level of society—which in turn lets the economy prosper.

The design, construction, commissioning, operation and subsequent operation of accelerator-based research infrastructures requires researchers from many different disciplines including physics, engineering and computer sciences to work closely together. Despite the need for skilled experts in this area, there are very universities in the world that offer structured courses on accelerator physics as part of their curriculum and often researchers have to be re-trained 'on the job' after their graduation to PhD in one of the above areas.

To overcome this gap in specific research skills training, the Innovative Training Network (ITN) scheme within the European Union's Marie Skłodowska Curie (MSCA) has provided unique support to the accelerator community for more than 10 years and sets one of the best examples for maximizing the investments in fundamental research. The scheme supports competitively selected research networks which combine partnerships of universities, research institutions, research infrastructures, businesses, SMEs, and other socio-economic actors from different countries across Europe and beyond. Each ITN enables cutting edge R&D and provides network-wide training to its Fellows during the 4 year project duration. Any subject area can apply and the best ideas are selected in a bottom-up approach.

ITNs exploit complementary competences of the participating organisations, and enable sharing of knowledge, networking activities, the organisation of workshops and conferences to train their Fellows which are usually employed by different host institutions for 36 months. With a success rate of only 5–7%, the ITN scheme is amongst the most competitive funding schemes.

The University of Liverpool/Cockcroft Institute has been exceptionally successful in coordinating ITNs in accelerator science. These programs started in 2007 with DITANET, a research network focusing on R&D into advanced beam diagnostic techniques for particle accelerators and light sources [7] which was proposed to the physics panel within MSCA in Framework Program (FP) 7. The network was an enormous success: With a funding of €4.2 million it trained 22 Fellows (PhD students and Postdocs), organized 4 international schools with up to 100 participants, as well as 9 international workshops for 30–70 participants, as well as an outreach symposium and final conference on beam diagnostics for the world-wide accelerator community. Presentations from all events remain accessible via the project home page and continue to serve as a unique knowledge base for the accelerator research community.

Building up on the successful collaborative model of DITANET, two further ITNs started within FP7 in 2011: oPAC (Optimization of Particle Accelerators) which was submitted to the physics panel and received €6 million to train 23 Fellows [8], making it one of the largest ITNs ever funded, and LA3NET (Laser Applications at Accelerators), submitted to the engineering panel which received €4.6 million to train 19 Fellows [9]. The two networks ran in parallel and shared a number of training events and also stimulated researcher exchange programs.

Taking the oPAC approach further, but focusing on a more specific area within accelerator science, OMA (Optimization of Medical Accelerators) was evaluated by the life science panel and was the first-ever ITN that received a 100% evaluation mark. OMA started 2016 to train 15 Fellows with a budget of €4 million and will run until the end of 2020 [10]. Finally, AVA (Accelerators Validating Antimatter research) started in 2017 and was selected by the physics panel to train 15 Fellows with a budget of again €4 million [11].

Group photo of AVA School on Antimatter Physics at CERN in Geneva, Switzerland. One of the MSCA ITN projects for next-generation of researchers that show a high multiplication factor for the impacts of investments in fundamental science

A structured combination of local and network-wide trainings is the central concept of all ITNs. Existing and well-proven training schemes are typically exploited, but at the same time novel training opportunities are made available which no single partner alone could offer. For example, hands-on training through research at accelerator facilities is a unique training opportunity which rarely can be provided within standard university doctoral programs.

With the exception of several Postdocs who were employed in DITANET (the option to employ Postdocs did no longer exist in later projects), most Fellows were registered for a PhD. This embeds them into a structured course program at their host university or, if their work contract is with an industry partner or a research center, with a collaborating university. Courses are selected at the start of their project in discussion with their supervisors, based on their project needs and their own background and reflected in their individual career development plan (CDP). In addition, network-wide trainings bring the Fellows together on a regular basis. This ensures that in addition to research-based overlap and links between projects, interpersonal links between the Fellows are established. Most network-wide events also include external participation and hence efficiently link the Fellows to the wider scientific community.

Within higher education there has been a move to provide graduates with the skills and knowledge required in society, equipping them for the world of work, often referred to as the 'skills agenda'. For example in the UK the development of transferable generic skills, in addition to those relating to subject disciplines, have been included in PhD research training [12]. However, such training is not formalized at many universities.

The ITNs mentioned above guarantee international competitiveness of the researchers trained within them by providing them with the necessary skills for a future career in either the academic sector or in industry. For that purpose, an interdisciplinary 5-day training program, designed for the particular needs of early stage researchers, is held in the first few months of each project at the University of Liverpool. This training is organized in collaboration with central university PGR teams, as well as key industry partners, including Fistral Consulting, Holdsworth Associates and Inventya. It consists of a 'project specific' part and a part addressing more 'general skills', which are based on group work.

This school programme was developed and tested during DITANET and has since been adopted as standard for all first year postgraduate students in the School of Physical Sciences at the University of Liverpool. This approach was praised during mid-term reviews of DITANET, oPAC and LA3NET and acknowledged as one of the 'best practices' in Europe. A final year advanced researcher skills training complements the general training. It focuses on the next career step and includes sessions on CV writing, interview skills, international networking, grant writing opportunities, technology transfer, and international career avenues for researchers.

In terms of the scientific training, each ITN usually organizes at least two international schools in their core R&D areas. These events are open to 70–100 participants and all course materials remain available via the respective webpages, which can be easily accessed through the project home page.

In addition, the networks have already organized dozens of targeted scientific workshops at venues across Europe, and there are many more in the planning. Each workshop lasts 2–3 days, is open for network members, as well as external participants, and focuses on expert topics within the respective network's scientific work packages.

Finally, towards the end of a project cycle, a network typically also organizes an international conference. These conferences include sessions on all R&D aspects within the respective network and highlight the research outcomes. Following the example of DITANET, oPAC and LA3NET joined forces and organized an international Symposium on Lasers and Accelerators for Science & Society took place on 26 June 2015 in the Liverpool Arena Convention Centre. With speakers including Professors Brian Cox (Manchester University), Grahame Blair (STFC), and Victor Malka (CNRS), the event was a sell-out with delegates comprising 100 researchers from across Europe and 150 local A-level students and teachers.

More recently, a joint Symposium on Accelerators for Science and Society between OMA and AVA, as well as the UK Centre for Doctoral Training on Big Data Science LIV.DAT [13] has copied this approach and was held at the Liverpool ACC in June 2019. It was preceded in March 2019, by another successful symposium "Particle Colliders—Accelerate Innovation" organized in Liverpool's Arena with the support of CERN and the EU-funded EuroCirCol project [14] accompanied by an Industry Innovation Workshop.

Thus far, almost 100 early stage researchers have been trained within MSCA networks coordinated by the author of this paper in the field of accelerator science, and probably hundred more through other networks. They have produced remarkable research results and trained an entire new generation of accelerator experts. Further collaborative projects on this basis have already emerged and have driven science and technology in this field. The training approach behind these initiatives has impacted very significantly on the world-wide accelerator community where several thousand researchers have already participated in one or several of the international schools, workshops and conferences. It has also served as an example for postgraduate training schemes outside of accelerator science and was commended as "best practice" by the EU as part of several formal project reviews.

The career development of fellows who were part of previous ITNs has been exceptionally good. This is certainly due to the high quality of researchers who were recruited in the first place, but their feedback has clearly indicated that the training, international networks, and cross-sector experiences they had access to, has boosted their skills and career prospects.

Despite such a considerable number of experts already trained in accelerator science, there remains a shortage of experts to drive R&D in accelerator science. Future initiatives will base their training model on the ideas presented in this paper as this has proven to provide maximum benefit for the Fellows, the R&D projects, and the institutions involved.

4 Summary and Outlook

R&D into particle accelerators has been driving innovation for more than 100 years. This has resulted in break-through scientific discoveries and enabled applications with enormous benefits for society. In the twenty-first century we clearly see a shift towards collaborative-driven research as discussed in the above examples.

This structural change in the way science is done should inform our thinking on how to improve the societal benefits from fundamental research and when designing future research infrastructures—that should facilitate and enable similar networks. Fundamental research provides a fruitful ground for training the next generation of researchers. Secure funding and opportunities for continued exchange of researchers and knowledge are needed to ensure an even brighter future.

Acknowledgement Research reported in this article has received funding from the European Union's FP7 and Horizon 2020 research and innovation programmes under Marie Skłodowska-Curie grant agreement Nos 215080, 289485, 675265 and the INFRADEV-1-2014-1 call under grant agreement 654305.

References

1. L. Rivkin, Plenary ECFA meeting, CERN (2018), https://bit.ly/2CuGm9a
2. N. Smith, "The biggest surprises always came from serendipitous discoveries", E & T magazine 14(9), 2019, https://bit.ly/34HQp6I
3. Science knows no borders, https://youtu.be/3pPAcrLUGX0
4. Busy bees and mighty magnets, https://youtu.be/lGImsJmwiXo
5. C.P. Welsch, "What have particle accelerators ever done for us?", Physics World (2019), https://bit.ly/2NxCRVE
6. Optimization of Medical Accelerators project, online brochure (2019), https://www.liverpool.ac.uk/oma-project/brochure/
7. DITANET, http://www.liv.ac.uk/ditanet
8. oPAC, http://www.opac-project.eu
9. LA3NET, http://www.la3net.eu
10. OMA, http://www.oma-project.eu
11. AVA, http://www.ava-project.eu
12. G. Wall, C.P. Welsch, "Employability in Europe: enhancing postgraduate complementary skills training", Proceeding of HEA STEM Learning and Teaching Conference Series, Birmingham, UK (2013).
13. LIV.DAT, http://www.livdat.org
14. https://indico.cern.ch/event/747618/timetable/?view=standard

Epilogue: Productive Collisions—Blue-Sky Science and Today's Innovations

Maria Sotiriou

I hope that readers who have got to this page will have found the present volume as exciting as I found the original workshop on the "Economics of Science" that the Belgium chapter of the London School of Economics (LSE) Alumni co-organised with CERN in Brussels in summer 2019. The LSE is today an established partner of the Future Circular Collider (FCC) Study, following a number of joint activities that the LSE Alumni Association Belgium co-hosted with CERN.

The contributors to this volume hail from the major European Big Science organisations. Academics working in economics and social sciences and representatives from funding agencies and EU institutions, brought alive, here in Brussels, the lasting and far-reaching impacts of Big Science. The workshop was a space for dialogue and exchange of best practices between these Big Science organisations, and I hope that these proceedings will inform the future design of research facilities, and boost their multiple impacts for society. These become evident just as we are seeing the final steps to launch Horizon Europe, one of the world's most ambitious funding programmes for research and innovation.

The collected essays convey the breadth of the arguments raised at our workshop, along with questions that we still face today as Europe prepares its new multiannual financial framework for innovation and research, including global scientific collaboration. I would like to thank the two editors of these proceedings as well as all the participants at the workshop, and of course Springer for publishing this volume in their series of Science Policy Reports.

As a co-host of the event, I was struck by the widespread sense that Big Science still today faces distrust from taxpayers and decision-makers alike. A sizeable bill that needs to be footed, the prospects of pay-off seem distant and there are oh so many more pressing budget needs.

M. Sotiriou (✉)
London School of Economics (LSE) Alumni Association Belgium, Brussels, Belgium

© The Author(s) 2021

H. P. Beck, P. Charitos (eds.), *The Economics of Big Science*, Science Policy Reports, https://doi.org/10.1007/978-3-030-52391-6_19

We are no strangers to such criticism here in Brussels. In answer, the contributions from the European Space Agency (ESA), CERN and the European Spallation Source (ESS) offer clear evidence that there are many immediate returns for society from the planning, construction and operation of such large-scale research infrastructures—before and beyond the pure scientific knowledge that can be often hard to quantify. Investing in blue-sky research pushes back the frontiers of our knowledge. But such investment also builds a broader ecosystem that includes industry, local communities, citizen scientists and the next generation of experts.

Big Science organisations such as ESA and CERN, are expert practitioners in international idea-pooling and idea-spreading. Fostering universal, inter-disciplinary approaches to discovery and sustaining distributed science work, Big Science organisations leverage their sites as places of citizen science—a place where everyone is welcome and everyone learns new—non-scientists and 'real' scientists, startuppers and students, young and old, people from many nations and backgrounds.

And contrary to their "elite" image in the popular imagination, we have learned how frequently transnational and international Big Science organisations contribute to justice and inclusion, offering opportunities to marginalised and often excluded groups. I find this theme—which emerges especially in the last four contributions to this volume—particularly intriguing, given the multiple crises that Europe has faced in the last years.

As you will have read in this volume, ESA has generated for its partner states socio-economic benefits that match investment Euro for Euro and more. CERN has demonstrated similar benefits: together with its collaborating institutes, laboratories and industrial partners, CERN has spearheaded huge breakthroughs, from the familiar information highway, the World Wide Web, to the very frontiers of human understanding of particle physics—who hasn't heard of the Higgs boson? These achievements stand as towering monuments to human inquisitiveness and ingenuity.

For those who continue to doubt the value of curiosity-driven investigations, the representatives from European Big Science organisations can also proudly point to spin-off innovations that bear fruit even while the Big Science investigations are underway, in the same way that space exploration and F1 racing technology make our everyday cars safer. One example would be the CERN contribution to proton therapy, a game-changer for some cancer treatments with particle beams and the development of isotopes for medical treatment.

These proceedings have demonstrated the long-term and copious R&D efforts needed to advance new technologies that will allow us to continue the exploration of nature at different scales: from satellite and space missions planned by ESA to next generation radio-telescopes like the Square Kilometre Array (SKA) Project, and from the upgrade of the Large Hadron Collider (LHC), and the planned FCC to spallation sources like the ESS, used for material science and biology.

In addition to research achievements, these projects also offer opportunities to strengthen European, and global, industry through joint co-innovation activities. They are also magnets to attract and develop the next generations of top-flight engineers, scientists and technicians. Fascination with the questions addressed by scientists working in these projects continually inspire young people and increase the attraction of science and technology more generally. Then, in turn, many of the young professionals trained at Big Science organisations transfer their expertise to other research projects, to industry and to society, at large.

Finally, and this is a particularly important remark for me as an LSE alumna, Big Science is a champion of global and peaceful collaboration. The CERN model offers today a powerful example of international cooperation driven by pure curiosity about our Universe, based on openness and inclusion. The LHC, today the world's largest scientific facility and largest user community, is the best proof of what humanity can achieve when acting together, and overcoming political and cultural barriers. CERN's history is closely intertwined with the European project: it remains the prototype for scientific collaboration in Europe. A global flagship hosted by Europe, well-known and well-established inspiration for sister organisations, ranging from astronomy to biology, and spinning out into fields such as the interplay between Big Science and Societal Justice: the theme of the last session of this workshop. I hope that in today's fractious climate, this volume will inspire continued commitment to global cooperation.

These are among the many reasons that I am proud of the results of the "Economics of Science" workshop reported here; an initiative that brought together the two driving forces that enable all this to happen: the funding agencies and Big Science.

We are lucky to have in this volume the voices of Big Science organisations carried by their people into the EU bubble and beyond.

Printed in the United States
by Baker & Taylor Publisher Services